A History of Earth's Biota

A History of Earth's Biota:

The Blooming of Life

By

J. William Schopf

Cambridge
Scholars
Publishing

A History of Earth's Biota: The Blooming of Life

By J. William Schopf

This book first published 2022

Cambridge Scholars Publishing

Lady Stephenson Library, Newcastle upon Tyne, NE6 2PA, UK

British Library Cataloguing in Publication Data
A catalogue record for this book is available from the British Library

ISBN (10): 1-5275-8724-X
ISBN (13): 978-1-5275-8724-3

Cover Image Buddhism's Sacred Lotus (*Nelumbo nucifera*), an aquatic plant having an underground stem and exceptionally long-living seeds that adapt it to a periodically wet and dry environment. *Nelumbo* seeds collected from a now-dry Lotus lakebed in Liaoning Province, northeastern China planted by Buddhist monks nearly two millennia ago, unearthed and germinated by the author's wife Jane Shen-Miller Schopf, have been radiocarbon dated at 1,300 years, the oldest directly dated viable seeds now known.

To my teachers, from whom I learned,
to my students, who teach me still,
and to my wife, who cheers me on.

TABLE OF CONTENTS

PREFACE

Background

As a second-year student at Oberlin College, a small undergraduate school in northeastern Ohio, I listened with interest to a stimulating lecture by Geology professor Larry DeMott during which he noted that the entire fossil record before the emergence of the oldest recognized animals (large many-legged trilobites) was *"unknown and unknowable"* – and had been for *"the past 100 years."* He then went on to note, almost in passing, that this was *"the greatest unsolved problem in all of natural science."*

As a "wet-behind-the-ears" young college student, this hit a resounding chord – but I was skeptical. That evening I searched through my paperback copy of Darwin's 1859 *Origin of Species* to find the answer. The prof was correct, Darwin penning that this "missing" early fossil record was *"inexplicable,* [an absence that could] *be truly urged as a valid argument against* [his theory of evolution].*"* Darwin was also correct – this <u>was</u> inexplicable – but I could not imagine why the problem had remained unsolved for a full century. This made no sense! After all, evolution is a fact – not some fanciful made-up "fake news" – and trilobites were obviously far too complex to have anything at all to do with life's beginnings. Darwin had tried mightily to explain the problem away (perhaps primitive life was too small, too fragile to be preserved, or perhaps all truly ancient fossil-hosting rocks had eroded away), but his notions about pre-trilobite life and the surviving rock record were by this time 100 years out of date.

Later that spring (1961) I journeyed off to Harvard to meet paleobotanist Elso S. Barghoorn who had written the one published paper I could find on the topic, a 1954 report lead-authored by economic geologist Stanley A. Tyler (University of Wisconsin) reporting what they regarded to be authentic pre-trilobite (Precambrian) ancient fossils. At the end of my visit, Barghoorn handed me a chunk of the black shiny rock (from the 1,900 million-year-old Gunflint Formation of southern Ontario, Canada) that Tyler and he claimed to harbor the tiny fossils. That was enormously kind of him a splendid souvenir? Or, perhaps, he wondered what can this young upstart do with this?

After checking the Harvard graduate-school entrance requirements (*"only Honors Students will be considered"*) I petitioned the Oberlin Geology faculty to allow me to do a research-based undergraduate Honors Thesis. Though the Chair of the Department demurred, arguing that "undergraduates have no business doing research," my petition succeeded. Following a faculty vote, I thus became the first Honors student in the department since the program had officially been put in place in 1908. Upon graduation in 1963, I trekked back to Harvard to become Barghoorn's student.

For me, this undergraduate research project turned out to be a real boon. When I arrived at Harvard I already had a couple years' experience doing hands-on research and – more than that – it enabled me to play a part in the first two 1965 "breakthrough" publications that laid the foundation for the field. The first of these was a decade-long-delayed full-blown description of the Gunflint fossils (the subject of my undergraduate research) co-authored by Barghoorn and Tyler (who had unexpectedly died in the fall of 1964). Its publication, however, engendered widespread skepticism, the naysayers opining that *"the 'fossils' are much too old, there must be some mistake!"* The second paper, published some six months later and co-authored by Barghoorn and me, reported a completely new find – diverse exquisitely well-preserved pre-trilobite, Precambrian fossils, this time from the 850 million-year-old Bitter Springs Formation of central Australia (which would later become the subject of my doctoral thesis). This second 1965 paper was the "deal-sealer" – different rocks, different continent, different age, and chock-full of abundant, varied, remarkably well-preserved microscopic fossils, many easily relatable to microbes living today. Taken together, these two seminal papers laid the groundwork for the soon-to-emerge and now thriving field of Precambrian paleobiology.

To me, today, those early years are a blur. I co-authored with Barghoorn a dozen or so scientific publications, gave numerous talks at scientific meetings and various universities, was elected to Harvard's elite Society of Fellows (like Tom Kuhn whose seminal studies are discussed in Chapter 8) and, upon graduation in 1968, landed a faculty position at the University of California, Los Angeles (UCLA). A pretty nifty beginning!

In the years soon after joining the UCLA faculty I continued to have enormously good fortune – youngest to be advanced to tenure (age 27), youngest Full Professor (age 31), and a series of awards from the University (for teaching, research, and academic excellence). My science prospered. In my early years, I journeyed off to Australia, India, the former Soviet Union, and China – in each, "spreading the word" about life's wondrous Precambrian fossil record and, importantly, in each country making friends and discovering the most ancient fossils there known.

To my great surprise, a "Biggie" then arrived. In 1976, the bicentennial of the founding of the United States, the US Science Board established the Alan T. Waterman Award to honor "one outstanding young scientist in United States" and a $150,000 prize (an honor now, decades later, that carries a prize of one million dollars). What was I to do with the wholly unexpected largesse, the "manna from heaven" that had landed on my plate? I pondered the matter for several weeks and came up with a plan – use the prize money to set up an international interdisciplinary team to work together on life's early history. I did. My notion succeeded.

In 1978 I assembled a group of 22 mostly then-young workers from Australia, Canada, Germany and the USA to come to my lab at UCLA, families in tow. Calling ourselves the Precambrian Paleobiology Research Group (PPRG) we worked together as a team for the following 14 months. We did our stuff and produced *Earth's Earliest Biosphere, Its Origin and Evolution* (J.W. Schopf, Ed., Princeton Univ. Press, 1983, 543 pp.), a volume that received that year's US national prize for a scholarly volume in its field. But we had covered only the first half, the earliest two billion years of Earth's history. *A job half-done is a job undone!* So, I then put together a second PPRG team to investigate the more recent two billion years of our planet's history – this time numbering 42 participants with new members from Denmark, Sweden, South Africa, and the USSR – which produced *The Proterozoic Biosphere, A Multidisciplinary Study* (J.W. Schopf and C. Klein, Eds., Cambridge University Press, 1992, 1348 pp.). It, too, received a national scholarly publishing award. Taken together, these two mammoth volumes – jocularly (if affectionately) referred to by some in the group as the "Old and New Testaments" – brought together what was then known about Earth's first four billion years of evolutionary history, the exceedingly long pre-trilobite Precambrian interrelated biotic-environmental history of the planet.

Given that spur, and in an effort to make this new knowledge accessible to a wider audience, I then wrote *Cradle of Life – the Discovery of Earth's Earliest Fossils* (J.W. Schopf, Princeton Univ. Press, 1999, 367 pp) to use in my Freshman-Sophomore General Education course, "Major Events in the History of Life." My Goodness! This volume, too, was a national prizewinner.

Why this book?

As is suggested in the prologue of my earlier effort, *Cradle of Life*, if one considers Earth's entire four-and-a-half-billion existence, Darwin's "missing" pre-trilobite Precambrian fossil record would encompass the earliest nearly 90%. Pause for a moment and compare that with America's almost 250-year-history. What that would mean is that today we would have no writings, no evidence, no facts and no way to understand anything about our country's history except for its most recent 25 years! Knowledge of all earlier events would have been wiped away – Benjamin Franklin, the Declaration of Independence and the Constitution; George Washington, Thomas Jefferson, Abraham Lincoln and the Civil War; electricity, telephones, radio and television; the Great Depression, two World Wars and the 45-year-long East-West "Cold War"; personal computers, the internet, mobile phones and much, much more. The biota-birthing saga that unfolded during that earliest 90% of life on Earth is the focus of *Cradle of Life*.

But what about the last 25 years of American history, the most recent 10% – don't they matter, too? Yes, of course. Global warming, the Great Recession, unemployment, gender equity, LGBTQ rights, political divisiveness, racial injustice, nation-wide protest demonstrations, widespread economic woes, two presidential impeachments, the covid-19 pandemic, the crisis in Ukraine and on and on and on – all game-changers, each occurring at a seemingly ever-quickening clip.

Interestingly, this latest phase of the American experience has a near-perfect parallel in the most recent 10% of life's long history – a segment of geological time referred to as the "Phanerozoic" (the "Age of Large Life") – and it, too, is of great interest, maybe even more so than life's Precambrian origin and enormously long, laborious, formative beginnings. And the parallel is not surprising – both Phanerozoic life on Earth and societal life in the United States evolve over time, each building on that which occurred before, each testing one possible advance then another and another, each moving sporadically, haltingly, as they search for the best available solution.

Moreover, it is only at the very end of this latest Phanerozoic 10% of life's history that we humans entered the scene, in terms of any such "geologic clock" just a scant few seconds before the present! What happened before humans finally emerged? Why, for what reason and in what order, did those stage-setters occur? Why do we humans walk on two legs, not three (a lot more stable in a windstorm), and why do we have only two arms, not three or perhaps five or six (a lot better to handle multiple objects at the same time)? And what is "intelligence" – a human trait that we love to tout – where in the world did *that* come from? Fortunately, the answer to such questions is simple and amply evidenced by fossil record-recorded Darwinian biological evolution. Indeed, as Theodosius Dobzhansky (a pioneering gene-studying fruit fly expert) taught us in 1973, *"Nothing in biology makes sense except in the light of Evolution."*

For the past two decades, I have used *Cradle of Life* as the prime text for my yearly Freshman-Sophomore General Education Course. Yet, as several students have remarked to me, *"it is not like any textbook I've ever seen it reads much more like a novel, an engaging and thought-provoking good story."* And that, again, is the aim of this book, a needed sequel because I well know that *Cradle* is lacking, covering only the earliest 90% of life's existence. The more recent 10%, the latest half-billion-years and that part of the record when plants and animals set the stage for humans to at long-last enter the scene, is what this book is all about. In short, *Blooming of Life* fills in the most recent 10% – to us, the most interesting phase of life's long history – aimed at linking together the highlights of the past half-billion-years of life's existence and showing how we humans are very much a result of life's evolutionary past.

How does the story proceed?

To begin this remarkable story – to get our juices primed to digest the tale that follows – Chapter 1 presents an overview of the history of life highlighting the two great world-changing advances of the pre-trilobite pre-Phanerozoic world and a short summary of how evolution works (a curtain-raising brief reprise of topics addressed in *Cradle*).

The narrative then moves on, oddly, you may think, the first two following chapters dealing with plants, not animals. Why should this be? After all, Darwin was concerned with the "missing" pre-trilobite (*pre-animal*) fossil record, not with the history of plant-life! The answer is simple. A great many animals, like us, are "land-lubbers" – ants, beetles, countless insects, frogs, lizards, birds, lions and tigers, virtually all the animals we know. But we, and they, could not exist on the land surface (or in the oceans, either) without plants. Plants provide the oxygen we breathe and, via that oxygen, the Earth's ozone layer that shields the surface of the land and ocean from the Sun's harmful UV-rays. Moreover, all of us – and all other animals as well – each and every day devour plants and (except for vegetarians) the meat of animals that fed on plants. Thus, today's entire ecosystem is wholly dependent on plants, even though we humans –

again rather oddly – tend to take plants for granted (as evidenced, for example, by the estimated 175 million Americans, roughly half the total population, who visit zoological parks and aquaria each year compared with the vastly smaller attendance at botanical gardens and arboretums).

In the following chapters, the narrative moves on to outline the history of animals – first sponges, jellyfish, worms, crabs and the other major groups of non-backboned animals; then fish (some that trundled, or at least slithered across the land), amphibians (like salamanders and frogs that make up the half-way tribe to full-blown land-animals), then reptiles (like turtles and dinosaurs) and birds (first flightless then ultimately soaring); and finally, mammals like us – a progression from early egg-layers (like the duck-billed platypus), to milk-producers with pouches to protect the new-born (like kangaroos), to placental mammals like us, those with a water-filled belly-sac to protect the unborn before they enter the world. A remarkable story! How, why and when did all that happen?

Blooming heads toward the climax with a next-to-the-last chapter that, using the perceptive insights of Thomas S. Kuhn, shows how science advances by using examples previously discussed in the text. The narrative then ties together the evolutionary story of the time-and-again rise, fall and ultimate success of life on Earth during the past half-billion-years of our planet's existence. Perhaps most interestingly, this saga features a series of surprisingly parallel innovations in plants and animals – from marshes, to the uplands, to the entire globe – with plant-life, the "eatees," leading the charge and animals, the "eaters" soon following their food. And though the volume does not include a discussion of the (very) recent rise of the human lineage – a rapidly evolving subject best left to the expert anthropology-archaeology community – this chapter closes with a short discussion of the origin of intelligence, a trait that some suppose is uniquely human but that in fact has exceedingly deep evolutionary roots, in its basics extending to the very origin of life on Earth. As you will discover, the most recent Phanerozoic 10% of life's long history is a fascinating tale!

The concluding chapter of *Blooming* then provides a surprising clincher, a fanciful yet instructive thought experiment that shows how exceedingly intermeshed Earth's plants, animals and environment actually are. The scenario is simple. Imagine that Earth is visited by benign, inquisitive, highly intelligent aliens. Imagine further that all of their analytical instruments have been disabled during their voyage though the Cosmos leaving them only the ability to collect and bring back to their home planet for study a representative "Noah's Arc" of Earth life. What could they learn? As surprising as it may seem, they in fact would be able to unravel a tremendous amount not only about today's living world but also about Earth's place in the Solar System, Earth's daily, monthly and yearly cycles, how Earth-life came into being, and how life and Earth's environment have co-evolved over the past 4.5 billion of years. Amazing stuff, all recorded just in the living organisms around us!

Thus, via a simple imaginative thought experiment, this final chapter illustrates the way science works, illuminating and tying together the central theme of the preceding chapters – the interrelated evolution and total dependence of life on its environment both now and over all of the geologic past – into a coherent whole.

Why is this book worth your time?

This is *not* a nitty-gritty "hard-science" reference-laden textbook. Rather, like *Cradle of Life* – its predecessor in this two-part sequence – *Blooming of Life* is intended for a non-specialist general audience, students and laypersons alike who are interested in learning about the Phanerozoic history of life, the 550-million-year-long history of plants and animals that set the stage for the rise of humans. In short, if you ever wondered about how this world around us came to be, where we came from and why it is that humans are "smart," this book will answer your questions.

The book is factual (at least as up-to-date and factual as I can muster) and the narrative is intended to be informal, readable, sprinkled with surprises and occasionally with a bit of humor (when I can find a proper excuse). To highlight the human side of science, it includes personal vignettes about several of the prominent scholars who have shaped current understanding of the topics covered, "movers and shakers" such as origin-of-life biochemist Aleksandr Oparin, microbiologist Carl Woese, micropaleontologist Boris Timofeev, fish expert Colin Patterson, vertebrate paleontologist Al Romer, global environmental authority Mohamed El-Ashry, and historian and philosopher of science Tom Kuhn. Moreover, as you read along you will find that it contains a fair amount about the historical development underlying the various concepts discussed, included to illuminate how scientific discoveries are actually made, and that it is well illustrated (at least the students seem to enjoy these pics). Though it does contain a few unavoidable "technical terms" (items for which there are no common names in general use), each such term is defined in normal parlance when it is first used and backed by a Glossary of Technical Terms. Moreover – and importantly, as you will discover – almost all of the fancy formal scientific names are also backed by an explanation of their linguistic derivation from ancient Greek or

Latin, an "add-on" not normally included in books of this type that you will find, as have I, is hugely helpful in understanding and remembering such seemingly odd arcane terms.

In short, *Blooming of Life* is intended to be a book written for *you* – presented in ways *you* can understand – an up-to-date summary of what is now known about the development of today's modern world as the volume wends its way through a terrifically interesting, remarkable story. Have a look at the Table of Contents. Browse through the images. Have a read and see what you think. I hope that you will be pleased!

CHAPTER 1

PRELUDE TO THE BLOOMING OF LIFE

When did dinosaurs appear?

Most of us know a bit about dinosaurs – those great beasts, some lumbering, some agile – that roamed the Earth in the distant past. Lucky grade-schoolers such as my older brother and I even got to see them, on Saturday mornings racing each other up the long stairs at Pittsburgh's Carnegie Museum of Natural History to see who could be first to reach the Dinosaur Hall. And many of us have enjoyed the 1993 movie *Jurassic Park* and its various sequels in which, I am pleased to report, the dinosaurs and their behavior seem remarkably accurate (largely and perhaps entirely because the producers of the film enticed Montana's expert on dinosaurs and their behavior, paleontologist John R. "Jack" Horner, to be their consultant). Yes, as depicted in the films, dinosaurs *did* lay eggs, protect their nests, and trundle about in great "flocks" … think of them as huge scaly flightless birds. Some were fearsome, like *Velociraptor* (a genus name that means "quick plunderer") and *Tyrannosaurus rex* (meaning "king of the tyrant lizards"), but others were more docile vegetarians, plant-eaters like *Triceratops* ("three-horned dinosaur") and *Brontosaurus* ("thunder lizard," now known by its original name, *Apatosaurus*, "deceptive lizard"). And yes, some of the meat-eaters, like the 'raptors, <u>did</u> hunt in packs, much like wolves and lions.

Like the great majority of life forms that once populated the Earth, dinosaurs are now extinct, having died-out long ago. Still, it is worth wondering when, in the long sweep of the history of this planet, dinosaurs actually existed. For most of us, that is not an easy question, largely because we think of time in human terms. One hundred years, a decidedly long lifetime, seems old. A couple of thousand years, back to the time of Christ, seems really ancient. And 450,000 years ago, the time of *Homo neanderthalensis*, our Neanderthal-forerunners with whom we share about 2% of our individual genes, is essentially unimaginable. But, in geologic not human terms, hundreds, or thousands, or even hundreds of thousands of years are trivial, piddling. Instead, Earth history is counted in *millions* of years, hundreds of millions of years, even thousands of millions (billions) of years. After all, planet Earth has been here for some 4,500 million (4.5 billion) years!

So, to put the question of "when did dinosaurs exist?" in more human terms, let's imagine that all of Earth history were condensed into the height of a man (**Fig. 1-1A**). Have a look and take a guess. At the kneecap? At the belly? At the chest? No, much later, at the forehead (**Fig. 1-1B**). In other words, as surprising as it may seem, dinosaurs are relatively recent!

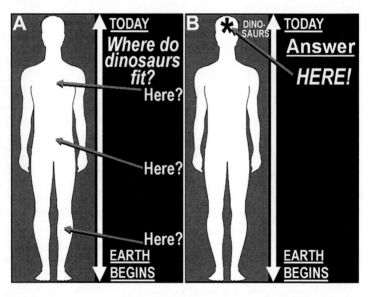

Fig. 1-1 (A) In comparison with the height of a human, when did dinosaurs appear in the history of the Earth? At the kneecap, the belly, the chest? **(B)** No, here, at the forehead! In the 4.5 billion history of the planet, the rise of dinosaurs was relatively recent, only a few hundreds of millions of years ago.

If that is so, what in the world happened earlier? The answer, now reasonably well understood, is an almost unimaginably long 4-billion-year-long series of events during the 4.5- to 0.5-billion year-long Precambrian Eon of Earth history that by about half-a-billion-years ago had finally set the stage for the rise of organisms that in their basic makeup, having heads, bodies, limbs, and so forth were like us.

Major advances of the Precambrian, pre-trilobite microbial world

Of all the vast number of evolutionary advances of the Precambrian pre-"life-like-like-us" world, two turned out to have monumental impact on the later history of life. Of these, the earlier – probably but not certainly as early as 3 billion years ago – was the advent of oxygen-producing photosynthesis (**Fig. 1-2**). Why did this matter?

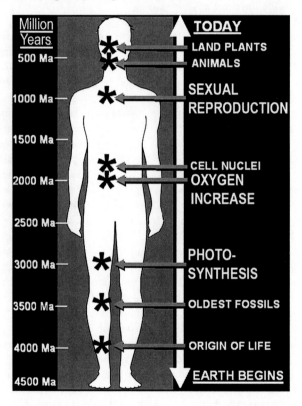

Fig. 1-2 Major events in the history of life, a great many dating from the earliest pre-animal Precambrian nearly 90% of Earth history. Among these game-changers, two stand out: The first, the microbial invention of oxygen-producing photosynthesis about 3,000 million (3 billion) years ago, with oxygen building-up to the amounts needed by oxygen-dependent life about a billion years later. The second, the origin of sexual reproduction about 1,000 million (1 billion) years ago, provided a huge advance for life that speeded evolution and life's ability to adapt to the ever-changing planet.

Life like us – all other animals and plants over the entire globe – relies on oxygen to breath. Think about it. That's the reason a flight attendant tells the passengers before a flight lifts-off that if the plane encounters problems and an oxygen mask drops from the ceiling, you should don it first and only then fix another mask to the youngster beside you. The attendant won't tell you why, but you have only about five minutes before you develop brain damage and another three or four before you are comatose. In other words, protect yourself first and only then "fight" with the little one at your side.

Our absolute requirement for oxygen is by no means a purely human trait. Indeed, all higher forms of life are wholly oxygen-dependent. Initially, however, when the planet formed, there was no oxygen to breathe. Where did it come from? Not from land plants – 4 billion years ago they did not exist. Rather, it originated in an early-evolved lineage of life – cyanobacteria, "pond scum," by far the most successful of all the truly ancient forms of life we know – microbes that could combine carbon dioxide with water to produce their life-sustaining glucose sugar and give off oxygen (in chemical parlance, $6CO_2 + 6H_2O \rightarrow C_6H_{12}O_6 + 6O_2$). In a short time, cyanobacteria took over the globe, chiefly as a result of "gas warfare," the oxygen they produced new to the environment and toxic to their competitors.

Nevertheless, it took a long while for environmental oxygen levels (now some 21% of the atmosphere) to build. Why? Mostly because oxygen is highly reactive, its nature being to combine chemically with other materials such as the gases issuing from volcanoes and the minerals of the Earth's crust. Because the planet had not seen this reactive gas before, the free oxygen produced by these microbes was sopped-up by the effluents of the volcanoes that peppered the early planet and by dissolved iron from deep within the Earth that belched forth from submarine fumaroles. When this iron welled-up from the ocean depths to reach the near-surface of shallow near-shore seas, it reacted in the uppermost thin veneer of oxygen-laden water to form tiny grains of the iron oxide mineral hematite (Fe_2O_3) that settled down to the sea bottom in a fine rusty rain. This then resulted in a long worldwide episode of the deposition of what are now known as "banded iron formations," mineable iron-rich rocks that form the basis of today's steel industry. Remarkably, the entire world rusted for hundreds of millions of years! Finally, by about 2.3 billion years ago, as the submarine sources of the iron slowly became depleted and early volcanism gradually abated, oxygen in the atmosphere began to build. Soon thereafter, oxygen-producing and -dependent algal plankton, single-celled "eukaryotes" having cells with nuclei as do we, entered the scene.

The second major game-changer before the evolution of many-celled higher forms of life like us was the advent of sexual reproduction about 1 billion years ago. Why sex? The answer is simple. Earlier life, both those with cell nuclei (single-celled planktonic eukaryotes) and those without nuclei (microbial cyanobacteria, for example) reproduced by cloning, each new generation having the same genetic instructions as the one before – the individual organisms died but their unaltered gene-based directives lived on generation to generation. However, with the advent of sex, the situation markedly changed. Sex combines the genes from two different parents – in humans from the mother and the father – the reason that offspring have traits from both and the source of what we call "family resemblance." Most of the two sets of parental genes handed down are pretty much the same, but there are enough differences that the offspring differ one from another.

Think about your brother and your sister. You and they are not exactly the same. Part of the difference comes from your genes, but another part from what you and they were taught as youngsters (an example of the continuous interplay between "nature," genetics, and "nurture," upbringing). Watch the kids in the checkout line at your local grocery store. Some children who are distraught will tug at their mother's skirt and call out *"mommy!"* while others will grab their father's shirt sleeve and holler *"daddy!"* Yet in some families, if the kids act-up they are likely to be scolded and told to *"act like adults."* Such variability illustrates aspects of the "nurture" side of the coin. And in most families, the first-born has a leg up on the siblings, a result of having received undiluted parental attention before the others entered the scene (leading to the 1- to 2-point slightly higher IQ test scores typical of the eldest offspring … "nurture," parental attention, here affecting the "nature," largely gene-based side of the equation).

Now, flip the coin over to its "nature" side. Genes, "nature," determine sexual characteristics and, even at a young age, the girls are likely to be more inquisitive, more imaginative and better with their hands – abundantly evident when grade-schoolers are introduced to cursive writing (a skill evidently no longer commonly taught). At a similar age, the boys typically are stronger, faster runners, more out-spoken, more rebellious. Such differences are of course not nearly as evident in genetically identical twins, especially in offspring of the same sex in which "nature" rather than "nurture" predominates. And this is true even for twins brought up in different families – they commonly like the same things, drink the same brews, wear similar clothes and share the same physical attributes as well as maladies – simply because their genes are identical as though they had been cloned.

Thus, such gene-based "natural" similarities among identical twins are not surprising. And because the precociousness of girls and comparatively laggard development of boys are also genetically determined, products of "nature" rather than "nurture," they too are easily explained. One of the most interesting and revealing studies on this subject, published in 2013 by Markus Kaiser and his colleagues at Newcastle University U.K., used brain-scans of a large population of young adults to document the development of their brains. What they found was that girls tend to optimize the nerve-connections inside their brains earlier than do boys, a ready explanation why females generally mature faster in certain cognitive and emotional areas than males during childhood and early adolescence. And all this, in turn, is a function of yet another genetically determine human trait, "puberty," when a child undergoes physical changes and becomes sexually mature, a process that typically begins around age 8 in girls and age 9 in boys. Indeed, the study found that girls begin to show greater brain-organization between the ages of 10 and 19 and that a comparable degree of brain maturation begins in boys between the ages of 15 and 21.

In sum, and regardless of how we humans have managed the situation, what this means is that the products of sexual reproduction, their offspring in every instance combining the "nature"-dependent genetic traits of two separate parental stocks, are bound to differ, at least slightly, one from another. Now, scale this up to the world's biota and compare the non-sex world with that after the advent of this new gene-mixing reproductive process.

Before life invented the process of sexual reproduction, the biota was fairly humdrum and monotonous, changing only very gradually – primitive microbes and even more advanced nucleated single-celled phytoplankton reproducing by the simple process of one cell dividing into two with their genes being passed along unaltered to the offspring – prompting some to refer to the billion-year lag between the origin of nucleated cells and the advent of sex as the *"boring billion."* Yes, life evolved, the ancient non-sexual bacterial microbes adjusting to exceedingly slow changes in day-length and solar luminosity and the later-evolved but similarly cloning single-celled nucleated phytoplankton gradually becoming a bit more diverse (with all such slow gradual changes being a result of random mutations of their genetic instructions).

Once sex arrived in nucleated cells, about 1 billion years ago, all this markedly changed – immediately, almost overnight. Why is that? Again, the answer is straightforward. By combining the gene-based traits of two different parental lineages their offspring also differed, not only from each other but from every other organism on Earth. In fact, you, me, and every one of us is unique – never before and never again will there be a human being precisely, exactly like a single one of us! And that it true as well about each organism of the entire sexually reproducing world. Thus, with the advent of sex, the diversity of living systems and their ability to adjust to and inhabit new environments immeasurably increased. In a flash, the speed of life's advance markedly accelerated. First known from the fossil record and now confirmed by gene-based phylogeny studies of living organisms, the results of this then newly devised method of reproduction is earliest shown about 1,000 million years ago by a rapidly accelerating increase in the varieties of single-celled nucleated plankton, protozoans and early-evolved multicelled algae. Within the next hundred million years this was followed by the proliferation of increasingly diverse many-celled seaweeds and then, later, by the forerunners of the many-celled animals and plants that, by 550 million years ago, began to populate the world.

Without the evolutionary inventions of early-evolved oxygen-producing photosynthesis, about 3 billion years ago, and the sexual reproduction of organisms having nucleated cells, about 1 billion years ago, we humans and all in the natural world around us would not exist.

Geologic Time Scale

Geologists, of course, do not think about the history of the Earth terms of the height of man. Rather, they and all of the scientific community uses the Geological Time Scale **(Fig. 1-3)** calibrated in millions of years (abbreviated as Ma, from Latin, *mega*, "very large, great" and *annum, "year"*") and stacked from the oldest to youngest, bottom-to-top, like a layer-cake, the oldest first-baked layer at the bottom of the pile and the later-baked layers laid one over the other.

The ages of the geologic units come from their radiometric dating, detailed studies of the time indicated by the natural alteration of one variety of a chemical element's inner structure to form a new, different atom or element. Atoms of the element carbon, for example, occur in three such forms, "isotopes," each denoted by its atomic weight – ^{14}C, ^{13}C, and ^{12}C – "carbon-14" being the heaviest, its extra weight caused by extra particles in its nucleus. Because of these extra pieces, ^{14}C is unstable – over time it "decays," comes apart by spontaneously losing the extra bits and changing into a completely new stable element, an atom of nitrogen, ^{14}N. This process, called "radioactive decay," occurs at a constant average steady rate. Thus, for example, in 5,730 years a pound (0.45 Kg) of ^{14}C will decay such that only half-a-pound remains, the lost half changing to ^{14}N, a length of time referred to as the ^{14}C isotope's "half-life." After an additional 5,730 years, only a quarter of a pound of the original ^{14}C will remain, a regular predicable process that goes on and on, the original amount of the "parent isotope" becoming smaller and smaller and the amount of the "daughter product" ever-increasing. Thus, once the isotope-characterizing half-life is known, all science has to do to figure out the age of the original carbon is to compare the amounts of the parent still remaining and the product it has produced.

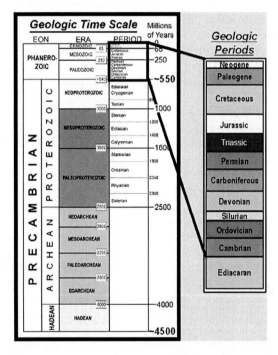

Fig. 1-3 Geological Time Scale. The total history of the Earth is divided into two great Eons, the earlier Precambrian Eon (from the formation of the planet, 4,500 million years ago, to the rise of many-celled animals 550 million years ago) and the later Phanerozoic, the Eon of large life, 550 million years ago to the present. Both the very latest Precambrian and the Phanerozoic are divided into Geological Periods, most named for the region in which its rocks were first studied.

This, of course, is the basis of the well-known carbon-14 method of dating, famous because of its use to date remains of early humans. But ^{14}C-dating has a serious drawback. As is true for all such radioactive isotopes, after about 10 half-lives – for ^{14}C some 57,300 years – too little of the parent isotope still remains for it to be accurately measured even by today's most sensitive instruments. Thus, carbon-14 dating works for only the past 60,000 years of human history, a paltry part of the human story which dates back to at least to Neanderthals, 450,000 years ago, and no doubt earlier.

How has science solved this problem? How is it that the 4,500 **million** years of the totality of the history of this world has been determined? In principle, the answer turned out to be simple – just use different naturally decaying elemental isotopes, particularly those having far-longer half-lives than ^{14}C. Here, the element uranium (U) which decays to lead (Pb) has especially come in handy, one of its isotopes having a half-life of 4.5 **billion** years, meaning that if a primeval chunk of Earth's original crust contained a pound of this isotope of uranium, only half-a-pound would still exist. And, as shown by U-Pb radiometric dating of the Moon rocks brought back by the Apollo astronauts in 1969, the Moon, formed at essentially the same time as Earth, is 4.5 billon years old (as are numerous meteorites, bits and chunks left over from the formation of the Solar System).

Because quite a number of such long-half-life isotopes have now been applied to the rocks of the Earth – measuring the isotopes in once-melted rocks, like volcanic lavas and thus showing the time when they congealed into solid minerals – the time-scale of Earth history is now well calibrated. Nevertheless, the high-tech instruments needed to make such measurements – mass spectrometers – are enormously expensive. For American colleges and universities this problem was solved in 1950 when, shortly after the end of World War II (an episode when _all_ Americans, academics included, participated in the war effort), President Harry S. Truman (1884-1972) established the US National Science Foundation (NSF) to support fundamental research and education in the non-medical fields of science and engineering. Over the preceding decades, the basic science underlying the dating of ancient rocks had become increasingly better established and, with the founding of the NSF, funding for the equipment needed to make the measurements became widely available.

So far, so good. But what does this have to do with life's long history? After all, isotopically datable once-molten rocks, like lavas, cannot contain fossilized organisms – any form of life, like all else in the vicinity of an erupting lava would have been fried to a crisp! Moreover, the fossil-evidenced history of the most recent half-billion-years of life's existence, the Phanerozoic "Age of Large Life," was already well known nearly a century earlier than the discovery of radioactivity in 1896 by the French physicist Henri Becquerel (1852-1908) and confirmed only two years later by the husband-wife team of Pierre Curie (1859-1906) and Marie Curie (1867-1934) by studies of the mineral pitchblende, the crystallized form of uranium oxide, UO_2 (discoveries for which Becquerel and both of the Curies shared the 1903 Nobel Prize in Physics). Of these luminaries, perhaps the most outstanding was Marie Curie (**Fig. 1-4**), a Polish and naturalized-French physicist and chemist who is the only person to win the Nobel Prize in two separate fields of science (Physics and in Chemistry). In 1906, she was also the first woman to be appointed as a professor at the University of Paris.

Fig. 1-4 Marie Curie, co-discoverer of radioactivity and the only person to receive the Nobel Prize in two separate fields of science.

Thus, in Darwin's time, the mid-1800s and a half-century earlier than the discovery of age-dating radioactivity, the basis for categorizing and systematically dividing-up the past half-a-billion years of the geologic column was provided by long-established knowledge of the fossil record, not by the radiometric dating of rocks. At first glance, this seems rather odd. Fossils, though sometimes almost whole and seemingly "life-like" are mostly just bits and pieces – shells, teeth, bones and the like – the resilient broken-up parts of the decayed bodies of dead organisms. How could such a messy mélange provide a reliable method for systematically divvying-up geological time – as in fact it did? And why did this development originate in England where it thus provided the basic underpinning for Darwin's theory of evolution?

In its essence, the development of this still-used system for ordering geological time was a result of the then-worldwide primacy of the British Empire and the Industrial Revolution it spawned – a good example of the interlacing of societal norms and the science it produces. British sailing ships collected raw materials from Britain's vast holdings – Australia, India, and large swaths of Africa and North and South America, nearly 25% of the land area and population of the world. The collected raw goods were then shipped to England to be converted into products to be sold back to the colonies.

The ships returned to their home port of Plymouth, at the southern edge of England, and because the manufacturing centers were to the north, in and around the Manchester area, an extensive canal system was built to connect the two. In the late1700s, William Smith (1769–1839), an English geologist credited with later creating the first detailed nationwide geological map of any country, served as a surveyor for the canal company. From this and his other geological work he became well acquainted with the strata country-wide and, importantly, the fossil faunas they harbored, the two together prompting him coin what he termed "The Principle of Faunal Succession," his observation that although the fossils changed from the lower (older) to upper (younger) rocks of a local sequence, the same faunas occurred place-to-place time and again in the same set of rocks throughout the country. Throughout his life, Smith continued to collect samples and map the locations of the strata he visited, thus amassing a large and valuable collection of fossils present in rocks he had personally examined – mostly preserved sea creatures such as trilobites, clams, brachiopods and numerous other types of "sea shells" – assembled not only from the sides of canals but from road and railway cuttings, and quarries and escarpments across the country.

This seminal work in England and adjacent Wales prompted systematic classification of the strata of the geological column and adoption of this system worldwide, a product – like England's extensive canal system – of the dominance of the British Empire. The Empire's hub, London, came to be regarded as the intellectual knowledge-rich center of the Western World. Even the "best and brightest" of colonial scholars – including those in Australia, Canada and the then-fledgling United Stares, for example – routinely sent samples of what they imagined to be major finds to London's Royal Society in the hope of obtaining affirmation from the experts. Thus, William Smith's fossil-based system for the ordering of strata in Britain had huge influence, particularly in adjacent Europe, while Darwin's proximity to London, his home in the nearby rolling hills ("Downs") of Kent, certified him as a serious scholar.

As a result, the major subdivisions of the past half-billion years of geological time, segments of Earth history known as "Geological Periods," were initially defined on the basis of the fossil faunas they contained – Smith's "Principle of Faunal Succession" – whereas the older underlying rocks, all lacking such fossils, were lumped together and mostly ignored. In ascending order, from oldest to youngest, these 11 Geological Periods are the Cambrian, Ordovician, Silurian, Devonian, Carboniferous, Permian, Triassic, Jurassic, Cretaceous, Paleogene and Neogene **(Fig. 1-3)**. To these, in 2004 a 12th Geological Period was added, the very-latest pre-Cambrian (pre-trilobite, pre-animals-with-hard-parts) Ediacaran Period, named for the Ediacara Hills of South Australia and an Australian Aboriginal name that denotes a place where water is or has been present, water being synonymous with the presence of life in the "Australian outback."

Taken together, the sequence from Cambrian to Neogene (the initially named 11 Geological Periods) comprises the ***Phanerozoic*** Eon (from ancient Greek meaning "visible" or "large life") – The Age of Large Life – an Eon that in turn is composed of three major subdivisions known as Eras. These Eras, in ascending order are the ***Paleozoic*** Era (from ancient Greek *palaió*, "old" plus *zōion* "animal") – The Age of Spore Plants and Marine Animals – composed of the Cambrian to the Permian Geological Periods. The immediately younger Era, the ***Mesozoic*** Era (from ancient Greek meaning "middle life") – The Age of Naked Seed Plants and Dinosaurs – includes the Triassic to the Cretaceous Periods. And the most recent of the three Eras, the ***Cenozoic*** Era (from the Greek phrase meaning "recent life") – The Age of Flowering Plants and Mammals – contains the Paleogene and Neogene Periods.

Taken together, the whole batch is a terrifically long list of names to remember– the Eon name (one) and the Era names (only three) being the easiest. But how can you recall all those other names in the right order for the 11 Geological Periods? If you are really good at memorizing lists you could simply remember them by the first letter of their names COSDCP, TJC, PN. But that is unpronounceable gibberish! And, of course, after you use them for a while they will become second nature. But to start out, a good trick is to make up some nonsensical sentence that, because you can turn it into an easy to remember image, will stick in your mind. For example, picture a lone cowboy sitting off in a corner, his hat pulled down and his hands covering his face, while the others in the room are having a roaring time as they raucously square dance. Now, with that image in mind, for the Paleozoic try "Cowboys Only Sit Don't Commonly Prance." For the Mesozoic, "They Just Cover." And for the Cenozoic. "Privately Napping." For all Periods together, "Cowboys Only Sit Don't Commonly Prance –They Just Cover – Privately Napping." Even better, make up your own easy-to-visualize mnemonic sentence, maybe using the names and traits of your friends and pals.

Where do all of these odd difficult-to-remember Period names come from? The answer is simple – virtually all come from the geographical name or the history of the region where the first well-studied sequence of the fossils and strata of the age occur. (If you want someone to blame, William Smith and his 18th Century contemporaries would lead the list.)

Here, in ascending order are the name-sources of the Phanerozoic Geological Periods and their ages in millions of years (Ma):

Cambrian (541-485 Ma), from *Cambria*, the Latin (Roman) name for Wales.

Ordovician (485-445 Ma), from *Ordovices*, the Latin name of an ancient tribe indigenous to North Wales.

Silurian (445-420 Ma), named after an ancient Welsh Celtic tribe known as the *Silures*.

Devonian (420-360 Ma), from the first-studied rocks in Devon (known also as Devonshire), a county in southwest England.

Carboniferous (360-300 Ma), from Latin *carbo*, "coal" plus *ferrous*, "producing, containing, bearing" – "coal-bearing" strata – named for the fossilized coal swamp floras of the central USA.

Permian (300-250 Ma), named for the Perm (Ural) Mountains near the city of Perm, Russia, 900 miles (1500 km) east of Moscow.

Triassic (250-205 Ma), named for the three-part division of rocks of this age in Germany.

Jurassic (205-150 Ma), named for the Jura Mountains, a sub-alpine range north of the Western Alps that parallels the France–Switzerland border.

Cretaceous (150-65 Ma), Latin for "chalk," named for the extensive beds of limestone found in the upper Cretaceous of Western Europe and southern England.

Paleogene (65-25 Ma), from ancient Greek *palaió*, "old, ancient" plus *gennaō*, "born" – "old born."

Neogene (25 Ma-present), from ancient Greek *néos*, "new, young, fresh" plus *gennaō*, "born" – "new born.

Why is a Chicken like a Man?

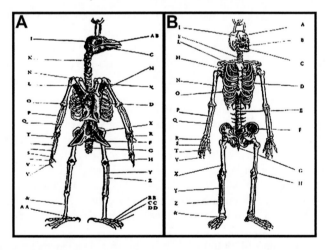

Fig. 1-5 A chicken is like a man! A drawing by French naturalist Pierre Belon that appeared in his 1555 volume *De La Nature Des Oyseaux* (On The Nature of Birds) illustrating and identifying the same sets of bones, albeit of differing sizes, in the skeletons of a chicken (**A**) and a human (**B**).

A chicken is like a man! Really? Have a look at **Fig. 1-5**, a drawing by French naturalist Pierre Belon (1517-1564) that appeared in 1555, more than 100 years before the founding of modern studies of comparative anatomy by English physician Edward Tyson (1651-1708) in the 1680s. In fact, the same suite of bones that Belon pictured in 1555 are present in <u>all</u> vertebrates ranging from humans to horses, walruses to whales, even in frogs and salamanders. Such different animals, yet their bodies and limbs are composed of the same set of bones! And even though their bones differ in size and shape, the fact that they share the very same set seems plenty remarkable.

Why should the bone-structure of all backboned vertebrates be essentially the same? Belon, like other European Renaissance scholars would have had a ready Biblical explanation, namely that God the Creator applied a single successful body plan to all such life, changing only bits and pieces from one to another to insure their success. Interestingly, as the German zoologist Ernst Haeckel (1834-1919) showed in his 1874 volume *Anthropogenie oder Entwicklungsgeschichte des Menschen* (Anthropogeny or Human History), such similarities extend even to the pre-birth fetal development of vertebrates – fish, amphibians (salamanders), reptiles (turtles), birds (chicken) and mammals (pig, cow, rabbit, human). Moreover, and perhaps even more surprising, some traits of the adults of animals that occupy lower links on the "the Great Chain of Being" (a hierarchy of nature that dates from pre-Renaissance Christianity) – for example, the gill slits that bring oxygen-rich water into mature fish – are present in the developing fetuses of animals higher-up the Great Chain, humans included (**Fig. 1-6**). Here, again, it seems truly odd that land-dwelling humans have gill slits at any stage in their lives. What good do they do?

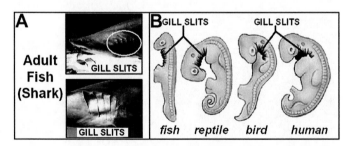

Fig. 1-6 Like adult fish (**A**), vertebrate embryos (**B**) have gill slits, a feature seemingly unnecessary in air-breathing reptiles, birds and humans that has been embedded in their fetal development by their common evolutionary ancestry.

Pause for a moment and think back to the prime adage taught us by Theodosius Dobzhansky (1900-1975), namely that *"Nothing in biology makes sense except in the light of Evolution."* Well, such similarities in bone structure and embryology are part-and-parcel of basic biology, so if

Dobzhansky is correct, evolution is the answer. And, indeed, it is! The evolutionary history of backboned animals, established by the fossil record of the past many hundred million years, shows that all are tied together in a continuous progression that leads from fish to amphibians, then reptiles, birds and mammals. We humans are no exception. We occupy only a tiny recent addition to this long-unfolding saga but we, too, are inextricably linked by our genetic composition – and thus by our anatomy and biochemistry – to <u>all</u> life that came before, bacteria included. By the process of evolution, past traits are modified and carried onward to the next lineage in the chain … Darwin's "descent with modification." This then explains the commonality of bone structure and even the shared presence of gill slits (which may seem worthless to humans, but were they to be subtracted from a developing embryo the fetus would immediately perish).

All that is true, but there is one small caveat to keep in mind. "Look-alike" traits, structures like the limb-bones of vertebrates (**Fig. 1-7**) for which the instructions are genetically carried forward from one lineage to the next, not only look alike but they also "do-alike" by performing essentially the same or similar function in each and every lineage. Such "look-alike-do-alikes" are therefore referred to as *homologous traits*, or more simply homologues (from the Latin *homos*, "same," plus *logos*, "relation"), features that share a common evolutionary ancestry (whether anatomical, like bones, or biochemical, like the chemical pathway we oxygen-dependent aerobes rely on to breathe). Such evolution-derived traits, referred to as being homologous ("homologues"), evidence the common gene-based ancestry of their related lineages.

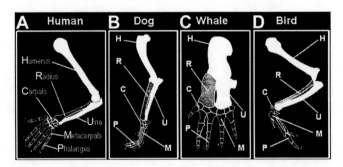

Fig. 1-7 "Look-alike traits," shown by the limb bones of a **(A)** human, **(B)** dog, **(C)** whale and **(D)** bird, share common evolutionary roots. In **A-D**, the shared bones, from the upper arm to the fingers, are identified by the first letter in their formal name (H, humerus; U, ulna; R, radius; C, carpel; M, metacarpal, P, phalanges).

Nevertheless, and although virtually all homologues are also do-alikes, not all "do-alikes" are homologous, not all features that perform the same or similar basic function are "look-alikes" that share a common evolutionary history. For example, compare a bird's wing with a butterfly wing (**Fig. 1-8**). Both wings do the same job – they enable their bearer to fly – so the two qualify as "do-alikes." But have a closer look. Bones support the airfoil of the bird's wing, the same set of limb-bones as that in all other backboned animals. But butterflies don't have such bones. Indeed, butterflies have no bones at all. Only lightweight veins support their wings – wings that are much smaller, much lighter than a bird's wing. And that's because butterflies are insects, members of a completely different, separate major tribe of animal life (technically, "protostomes") than vertebrates ("deuterostomes"). Even though their wings do the essentially the job as those of a bird or a bat, because of their differing evolutionary histories such features are referred as *analogous traits*, or more simply analogues (from the Latin *analogia, "pro*portionate," plus *logos*, "relation"). Although such "do-alike" but not "look-alike" traits are not widespread among the various animal or plant lineages, belonging to a small subset of features that illustrate what biologists refer as "parallel evolution," the distinction between the two matters. Homologues reflect evolutionary relationships, analogues do not.

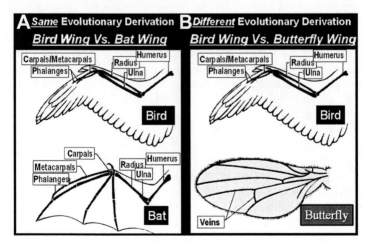

Fig. 1-8 Most "look-alike traits," homologues **(A)** such as a bird's wing and a bat's wing, are also "do-alikes" that enable their bearers to perform the same task, in this case to fly. But some "do-alikes," such as **(B)** a bird's wing and butterfly wing have different evolutionary roots – and are therefore referred to as analogous ("analogues") – that do not reflect a shared evolutionary history.

The roots of Darwin's theory

Given the foregoing introduction to the long history of life on Earth, a cardinal question remains, namely, how does evolution, the overriding concept that brings together all forms of life, actually work? Thankfully, as you will see, the concept of evolution is easy to understand. But the story of its genesis and final acceptance is a bit more convoluted – and more interesting – than most of us might imagine.

Presentation of the theory of evolution, of course, dates from 1859 and the publication of *On the Origin of Species*, the world-changing opus of Charles Robert Darwin (1809-1882). And many know that this volume was a product of his epical 1831-1836 world-encircling voyage as the naturalist on the *H.M.S. Beagle* (for which, with the approval of his wealthy physician father, he received no pay). What is not widely known is that his all-embracing concept, "Darwinian Evolution," was not entirely new nor was he its sole discoverer. The following is an abridged version of this fascinating little-known story.

In the first place, Darwin's concept had important precursors that played a big role in influencing Darwin's thought. For him, the first was almost certainly his grandfather, physician Erasmus Darwin (1731-1802) – whom Darwin never met, Grandpa Erasmus having died seven years before Darwin was born. Still, Darwin would have read his grandfather's 1794-1796 multi-volume medical volume *Zoonomia* that outlines the basics of the evolutionary progression. Only ten years later – in 1809, the year of Darwin's birth – the French naturalist Jean-Baptiste Lamarck (1724-1829) published *Philosophie Zoologique,* the first coherent simple-to-complex theory of what evolution produced (but a thesis flawed because his mechanism for the process, the use and disuse of life-improving traits that he imagined to be spurred by some unexplained "vital force" toward perfection, lacked backing). Then, in 1810-1820, another French naturalist, Baron Georges Cuvier (1769-1832), Napoleon's Minister of Education, based on his studies of fossils of the Paris Basin discovered extinctions. Like virtually all of his European contemporaries, Cuvier was a confirmed Christian Creationist who disputed Lamarck's claims. So, he interpreted his new find in terms of the Bible, in his view a record of God's repeated catastrophic extinctions followed by successive new creations.

We now come to the oft-neglected "mover and shaker" Thomas Robert Malthus (1766-1834) who 1798 published a hugely influential volume, *An Essay on the Principle of Population*. Malthus, a cleric in the Church of England, had the idea that human population, unless checked, increased faster than the ability of a country to feed it. To Malthus this pressure explained the existence of poverty, which he justified theologically as a needed force toward human self-improvement and self-control. Two centuries earlier, in 1601, the Elizabethan Poor Law had been enacted in England and Wales, rules that required the poor to be set to work, idle poor and vagrants to be incarcerated, and pauper children to become wards of the state or apprentices to the working class. But the new Malthusian theory of population made a strong and immediate impact on British social policy. It had been long believed that fertility itself added to national wealth by producing an ever-expanding workforce, but by the early 1800's many had become convinced that the 1601 Poor Laws had encouraged an increase in the number of over-sized poverty-stricken families and, as a result, an out-pouring of undeserved doles that were a drain on state coffers. Spurred by Malthus' views, enactment in 1834 of The Poor Law Amendment Act, known widely as the New Poor Law, changed all this.

In light of Malthus' national influence, it is not the least surprising that in 1838, "for amusement" according to his diaries, Darwin read Malthus' 1798 volume. Even though Malthus had written only about human populations, Darwin saw that the Malthusian principle of "want generating competition" was applicable to all life on Earth, Darwin's central idea being that life evolves in competition for the limited resources available and that over time this would result in a "Natural Selection" of the successful winners, the losers being winnowed out. Both in his diaries and in *The Origin*, Darwin indicates that he was well acquainted with the ideas of Lamarck, Cuvier and Malthus, and by 1844 he had completed his later-to-be-published volume, originally entitled *An Abstract on the Origin of Species and Varieties through Natural Selection*.

Even though Darwin's two mentors – geologist Charles Lyell (1797-1875), famed for his adage *"the present is the key to the past,"* and botanist Joseph Dalton Hooker (1817-1911), later to become Director of the Kew Royal Gardens on the outskirts of London – urged Darwin to publish his theory, he held-off (evidently to assuage his wife, a member of the prominent Wedgwood family and a devout Anglican) presumably imagining that it would be published posthumously when he would not have to deal with the aftermath. This, however, was not to be. In the spring of 1858, an unforeseen bombshell landed on Darwin's plate. Alfred Russel Wallace (1823-1913), a British professional specimen-collector then stationed in Indonesia sent Darwin a letter that enclosed a short manuscript that encapsulated the same ideas as Darwin's opus. Having delayed publication of his theory for now more than a decade, Darwin was about to be scooped! As it turned out, Wallace, too, had been spurred by

Malthus' book which he had read about 1846, a dozen years earlier, while
he lay in bed recovering from fever in Malaysia.

What was Darwin to do? He turned to Lyell and Hooker who
then arranged for both Darwin's and Wallace's ideas to be presented
publicly, back-to-back, at the July 1 1858 meeting of the Linnean Society
of London. Their ideas were presented, but they initially excited little
interest. Indeed, it was not until the following spring when *The Origin* was
finally published that the brouhaha erupted.

What happened next? Darwin's book came out. Its nature-
centered, rather than "theologically correct" ideas flew in the face of
accepted Biblical truth. Its impact was immediate, searing and far-
reaching. Indeed, only six years later, in 1865, Darwin's friend and
commander of the *H.M.S. Beagle*, Captain and later Vice-Admiral Robert
Fitz-Roy (1805-1865) committed suicide, evidently a result of his deeply
held religious beliefs that led him to be distraught for having "*helped
Darwin to disprove the Bible.*"

Soon after *The Origin* appeared, Darwin and his new notions
were mocked and derided in the press. Like it or not, such reactions to any
major game-changer – as Darwinian Evolution surely was – are par for the
course. As discussed in **Chapter 8**, they fit hand-in-glove with what
philosopher and historian of science Thomas S. Kuhn (1922-1996) has
called the "crisis phase" that typically follows the unveiling of any new
major concept-altering paradigm, especially one like Darwinian Evolution
that challenged the theologically entrenched prevailing view.

How does evolution work?

In its essence, evolution is simple, easy-to-understand. The "First Law of
Biology" is, obviously, to stay alive. All organisms are experts at this, well
fit (well adapted) to their local setting. By their own volition none would
change, none would evolve, if they didn't have to. Indeed, even if chance
mutations occur that alter their gene-coded instructions, they have a
multitude of biochemical mechanisms to fix the genes back to their
original unaltered condition (humans at least five, and evidently more as
well). To stay alive, life needs only to succeed in competition with others
in the immediate surroundings. For animals, this means that they compete
for food, an almost always limited resource, and to reproduce their kind,
those producing the most offspring – typically parented by the strongest,
fastest most "dominant" members of the group – winning the competition
as their genes flood the population. For plants it is much the same except
their "food" is light, needed for their life-sustaining photosynthesis that

produces the glucose sugar on which they depend. Thus, plants compete for space to soak-up sunlight and, like animals, they also compete to multiply their stock, the most successful taking over the pond or the field or the forest or your home garden (weeds, unwelcome "volunteers," and sprouts of bamboo spread underground from some nearby handsome patch, being prime examples).

If the environment does not change, all is well – a more or less stable equilibrium sets in among and between the animal "eaters" and their fodder, the plant "eatees." Nevertheless, the environment <u>does</u> change, planet-wide changes occurring slowly over millions of years – vastly more rapid, human-caused Global Warming being an obvious exception – whereas local changes typically occur in fits and starts over tens or hundreds of years. In fact, there are hardly any essentially unchanging environments on Earth, except the below-surface sub-seafloor mud that coats the miles-deep ocean abyss, a setting where there is no oxygen, no light and no wave action that has remained more-or-less as it is today since the early history of the planet. But there are almost no forms of life that live in such mud – most notably, early-evolved anaerobic (non-oxygen-requiring) sulfur-cycling microbes that have thus far been discovered in only one such deposit of the world's entire fossil record.

In contrast, virtually all life as we know it lives near the surface of the planet, on land, in the shallow ocean water that laps-up at the seashore, or floating and swimming in the upper reaches of the oceans. This makes good sense. Plants need light to photosynthesize and produce their life-sustaining glucose sugar, and all advanced forms of life depend on the oxygen they produce by harvesting that sunlight. Thus, to survive and produce that oxygen, plants, seaweeds included, must live within the near surface "photic-zone" where light is available. And because animals depend on plants – not only for the oxygen they produce but as a food-source as well – animals, simply to stay alive, live where the plants are. This cohabitation has existed for hundreds of millions of years as time-and-again plant-life colonized a new habitat and the animals followed in their wake, the "eatees" leading the charge with the "eaters" trailing along (and a repeated sequence that explains a lot about the whys and wherefores of how animals evolved).

Moreover, this understandable plant-animal occupancy of the same near-surface setting, spurred by the fundamental requirement to stay alive, is the root cause of Darwinian evolution. Phrased in its simplest form, *"evolution is the result of organisms competing within and interacting with and adapting to a changing biological-physical environment."* The near-surface environment provides just such a setting, changing almost daily. The air above is subject to constantly changing temperatures and dry-spells followed by windstorms, tornadoes, hurricanes and the like; the land surface is continually recast by flooding, erosion, even volcanic uplifts and glaciations; and the oceans, with their waves, currents and constant mixing are far from static. Moreover, and of particular importance to evolution, over geologic time the entire planet has changed, alternating back-and-forth between periods of dry conditions, as sea-levels dropped and the near-shore seas receded, to monsoonal times when sea-levels rose and near-shore basins flooded; from periods of intense cold and continent-wide glaciations to warmer more clement times, such as today; and from episodes of widespread volcanism and the uplift of new land to times when the continental landmasses remained relatively stable. Note that "geologic time," measured in tens, hundreds or thousands of *millions* of years, differs vastly from "biologic time" – calibrated in terms of organismal life-times and the number of generations required for marked genetic change – only *tens to hundreds* of years.

Such changes in the environment, coupled with the competition required to stay alive, provide the bases of Darwin's theory. Fueled by the additive short-term generation-to-generation biotic changes sparked by competition in the ever-changing environment and placing these on the unimaginably longer tapestry provided by geological time, Darwinian Evolution addresses the entire history of life over millions and billions of years.

Here are the rules of how evolution proceeds:

(1) Genes, the genetic instructions housed in chromosomes and derived from parents, dictate the form, function and life-history of an organism.

(2) Both the genes and the immediate local habitats vary, place-to-place, time-to-time.

(3) Such variation promotes competition, most broadly for food or light and to reproduce.

(4) In this continuous competition, the least fit (least adapted) and their "bad genes" are winnowed-out whereas the most fit (best-adapted) win and their "good genes" are passed along to offspring.

(5) As a result of these ever-changing conditions, the mix of life forms changes over time – some become extinct as other groups replace them and rise to dominance.

Notably, however, this evolutionary process has not resulted in a smooth steady progression of advance. Rather, time and again biological evolution has moved sporadically, marked by episodes of the appearance of new body plans and rapid bursts of the evolution of new forms of life, new ways to cope with changing conditions, separated by long periods of little or no change in the successful biota, prolonged episodes of evolutionary "stasis."

Life's evolution thus has a parallel – though on a decidedly different time-scale – in the more familiar brief ~250-year history of the "American experiment." As the newly formed union of disparate states moved from the courageous then-novel notions of the nation's Founders, it has experienced repeated episodes of unrest as the populace sought to improve conditions that did not match their expectations, and then back to relative stability – a back-and-forth history of upheaval followed by relative quiescence.

The same is true about life's history as documented in the fossil record. A bit more than half-a-billion years ago, well after seaweeds had emerged, mobile animals, all marine, appeared and within a scant few million years a great many new lineages entered the scene, an episode known as the "Cambrian Explosion of Life." After Earth's biota had adjusted to this "new normal," it settled down. Then, less than one hundred million years later, another great breakthrough occurred, the rise of plant life on land. With this new advance, food was available on the near-shore marshy land-surface – and animals (amphibians, like salamanders and frogs) soon followed. Then, after another tens of millions of years, plants developed seeds, allowing them to occupy the previously empty highlands, an advance later followed by the rise of land-animals with eggs (reptiles) and an advance that allowed the "eaters" yet again to follow their life-sustaining "eatees." And the story goes on and on throughout the history of life: A biota-changing major advance promoted upheaval of the world's ecosystem – a time of rapid evolutionary change as new opportunities opened-up – an episode that was then followed by a period of relative stability as Earth's biota adjusted to the new conditions.

In 1972, this on-and-off progression of Phanerozoic life was formally dubbed "punctuated equilibrium" by Stephen Jay Gould (1941-2002, Harvard) and Niles Eldredge (American Museum of Natural History), and was later (2007) popularized by Gould. To them and the general public this may have seemed a new concept, but it was not so to the paleozoological and paleobotanical communities who had documented and written about it for many preceding decades. That is the way of science … "new ideas," like even Darwinian Evolution, are quite often not entirely novel.

Evolution gives rise to myriad separate species

We all know that humans belong to the genus *Homo* (the Latin word for "man, human") and the species *sapiens* (Latin for "wise"). All humans belong to this single species, *Homo sapiens,* and *all* humans on Earth – regardless of their parentage, gender, country of origin, body build, skin color, religion, culture or anything else – are basically, fundamentally identical, and not merely in their needs, attitudes and aspirations. This is because although there are more than a million differences between your particular set of genes and those harbored by anyone else – differences that make each of us one-of-a-kind – we nevertheless share 99.9% of our total toolbox of genes with all other humans on the planet. And the same is true for all the myriad advanced species of plant and animal life in the world, from seaweeds to trees, from microbes to man, each member of each species being genetically essentially identical to all the others of that particular species, and each species being a separate biological entity unto itself, distinct in crucial ways from all other such species on the planet.

Moreover, like *Homo sapiens*, every species carries its own double-barreled genus and species name, typically derived from Latin or ancient Greek, the genus name also embracing other closely related organisms (rather like your own family's last name, Jones, Vu, Garcia or Shen) and the species name denoting unique stand-alone types (like your official first name, Miguel, Jane, Victor or Carola). In biology, the genus-species "binomial system" of nomenclature was devised by Swedish naturalist Carolus Linnaeus (Carl von Linné, 1707-1778) who introduced it in his 1735 volume *Systema Naturae.* Linnaeus' system of classification soon caught-on and in 1788, ten years after his death, the world's oldest still-active biological society, the Linnean Society of London was established and named in his honor (where, since 1829, it has served as the repository of his botanical, zoological and library collections).

If you pause for a moment and think about it, not only are the products of the evolutionary process innumerable, diverse, and each in its own way "special," but the whole structure of the biological world seems truly remarkable. How is it that life's myriad species have resulted from the rather simple process of Darwinian evolution, and how has each such species remained distinctly different from all others?

Fortunately, like evolution itself, this too is easy to understand. The entire process is based on genetics, the crucial chemical instructions housed in the genes of chromosomes. Thus, a biological species is defined as being composed of *"naturally occurring populations, the members of which are able to interbreed and are reproductively isolated from all other such populations"*. Thus, the three crucial species-defining terms are "population" – organisms of one kind living in the same area; "interbreed" – able to mate and produce viable offspring; and "reproductively isolated" – meaning that they can mate only with their own kind, not with other species. The first two of these defining characteristics, "populations" and "interbreed," are self-explanatory. But what about "reproductively isolated" – what does that mean and how is it maintained?

This, too, has a ready explanation. As Nature would have it, there turn out to be a huge number of genetically defined traits, both behavioral and environmental, that separate one species from another. Consider plants, for example. Some types come into flower at one particular time of the year while others flower at a different time. If they are not on the same time-schedule they cannot reproduce because their offspring-generating sperm and eggs are not available at the same part of the (usually spring) season. Likewise, some plants live in marshy settings, some high on mountains, still others on isolated oceanic islands – again, if they do not live in the same setting they cannot interact and co-produce. The list goes on – aridity, salinity, day-length, temperature, differing insect-pollinators, differing apparatus to accept sperm-carrying pollen grains, and so forth. Plants have a host of mechanisms to maintain the reproductive isolation of their individual species.

What about animals? Again, a plethora of isolating-mechanisms exist. Like plants (and all of the other varieties of life), animals cannot co-produce unless they live in the same environment at the same time and the genetics of the two potential parental stocks are closely similar, having the same number of chromosomes and virtually the same, and therefore compatible genetic make-ups. But, just how similar need they be?

To answer this question, consider chimpanzees (*Pan troglodytes,* the genus name *Pan* from Greek meaning "all-inclusive," as in the now all-too-familiar term **pan**demic; and the species name *troglodytes* from the Greek *trōglē*, meaning a cave or hole in the ground and here used to suggest that chimpanzees are like "cave men"). And think also about bonobos, so-called pigmy chimpanzees (*Pan paniscus,* the species name from Latin *pāniscus* meaning small chimpanzee). These two species are our closest living relatives with whom for chimpanzees we share about 99% of our genes and for bonobos only slightly less, 98.7%, a near-identity that has led some to refer jocularly to humans as "hairless chimpanzees." Is this striking genetic similarity sufficient for chimpanzees and humans to mate and produce viable offspring? Almost certainly not, but I cannot imagine any of us wanting to give it a try!

A close follow-on to the genetically mandated "First Law of Biology," to stay alive, is the gene-dictated "Second Law of Biology," to reproduce your kind. Why don't different species mate and produce hybrids – a mix that might be more successful than either of the parental stocks? In short, how is species-defining reproductive isolation maintained? In part this comes from the chemical incompatibly of the sperm cells of one lineage and the egg cells of another – the sperm of one species, for example, requiring alkaline conditions and the egg-cells of the other needing a more acidic setting, or the eggs of one stock lacking the biochemical signals needed to stimulate the frenzied onslaught of sperms of another needed to invade the unfertilized egg. Mostly, however, it is due to the genetically programmed behavioral cues that promote organism-to-organism interactions – signals that within a given species promote male-female interplay, a prelude to mating, but have little or no affect on other species.

How does such male-female interplay affect us humans? Well, consider why it is that young women commonly "dress-up" and put on lipstick before they go out on a date – in their words, *"to make themselves more attractive,"* that is, prettier and more likeable. And why do many soon-to-gradate high school girls practice and practice the difficult skill of walking in high-heels? Again, the aim is to make themselves "more presentable" as they carefully trundle across the stage to receive their diplomas. Males are much the same. Why is it that some high school boys tend to be loud and boisterous? They're showing-off for the girls, hoping to attract their attention. And why do most adult males shave their faces each morning? It's a bother, of course, yet the answer is simple, to remove the gene-dictated scruffy facial hair that would make them look older in an effort to instead retain a younger "more attractive" youthful look. The underlying causes of these gender-correlated interactions are complex, varying person to person and group to group. As a whole, however, they most probably are a mix of both gene-based "nature" and familial, societal, and peer-group influenced "nurture," a mix that quite obviously plays an important role in the way we humans develop and mature.

Finally, why is it that such boy-girl interactions rise to prominence so relatively early in life, in the teens rather than the twenties when mate-hunting becomes more serious? Here, the answer is that it is genetically pre-programmed. Indeed, as boys go through puberty and girls first experience their monthly menstrual cycles, their genetically programmed hormones kick in and both are capable of mating and producing children. To illustrate the point, go back in time a couple million years ago to *Homo habilis* (from Latin meaning "able man" or "skillful man") – a now long-lost ancestor of our species – and you will find from their surviving remains that the great majority died in their late teens. For *H. habilis*, a 35-year-old would be the exceptionally elderly leader of the group, an age when even today childbirth becomes an iffy proposition to both the mother and the fetus. Thus, for *H. habilis* the normal practice was to reproduce during their early teen-age years, a possibility available to modern humans as well though it is now frowned on by societal norms.

Fig. 1-9 A male turkey puffs his chest and displays his plumage in hope of attracting a mate.

Most all other animals are like us, the males showing off and calling out in their search to mate, the females (the males hope, trust) enticed by their advances. Consider, for example, a wild turkey (*Meleagris gallopavo,* a 1758 Linnaeus moniker from the Greek *meleágris* meaning "turkey, guinea fowl" and the species name a combination of *gallus*, "chicken," and *pāvō*, "peacock"). Though you might not imagine it when you excitedly view a roasted, featherless, stripped-down turkey at your yearly Thanksgiving Day Dinner, their mate-attracting plumage is spectacular (**Fig. 1-9**). Or, imagine a strutting peacock (*Pavo cristatus,* the species name from Latin *cristatus*, "plumed, tufted, crested"). Like a great many other birds, peacocks, in competition for mates with others of their species are showing-off … "Gertrude, here I am – come visit!" Though to us, their displays are stunning, beautiful – and their plumage *is* effective in attracting the mating-partners they are seeking – to other birds, even of similar size, it matters not at all. Indeed, the others either pay no heed or, in some such encounters the displaying males incite panic in the observers as the First Law of Biology kicks in and they scurry for cover to avoid being devoured by this enormously large creature.

The animal world is replete with many such examples, such gene-dictated species-isolating traits spanning animal behavior across all the various lineages. And they work! Each and every species remains distinct, "reproductively isolated" from all the others.

Life like us has cells like ours

As shown in **Fig. 1-10**, all organisms on Earth are composed of only two types of cells, those of the early-evolved microbial Bacteria and Archaea – cells known as "prokaryotic" (from the Latin *pro*, "before," and "*karyon*," a word from ancient Greek meaning "nut or kernel" used in biology to refer to the cell nucleus). All other forms of life are known as "eukaryotes" (*eu*, Greek, meaning "true, genuine"). Thus, prokaryotes are "before the nucleus" and eukaryotes are "truly nucleated." Superficially the two types of cells differ, but they actually are fundamentally similar, eukaryotes having evolved from prokaryotic ancestors and, therefore, incorporating and building on the traits of their evolutionary ancestors.

Compared one to another, a prokaryote cell (**Fig. 1-10A**) is the simpler – it lacks the bells and whistles of its evolutionary descendents – no cell nucleus and hardly any other added-on intracellular features. The more advanced eukaryotic cell (**Fig. 1-10B**), however, contains quite a bunch of separated, job-defined and membrane-enclosed individual intracellular factories (known as organelles, "tiny organs"). Included among these are *nuclei*, bodies that house the chromosome gene-carrying DNA- (deoxyribonucleic acid-) encoded instructions that via code-reading molecules of mRNA (messenger ribonucleic acid) are parceled out to *ribosomes* (tiny bodies, some 250,00 in each human cell) that produce the huge number of enzymes that midwife the chemical reactions necessary for the organism to stay alive. The third such type of organelle also present in all eukaryotic cells are *mitochondria*, the energy-producing factories that provide the chemical apparatus that enables their high-energy-yielding aerobic respiration, "breathing." Each of these eukaryotic organelles is enclosed by a flexible thin covering and they are thus referred to as being "membrane-bound." Unlike animals, the cells of eukaryotic plants (**Fig. 1-10B**) also contain membrane-bound organelles known as *chloroplasts*, chlorophyll-filled sacs that power their life-sustaining and glucose- and oxygen-producing photosynthesis.

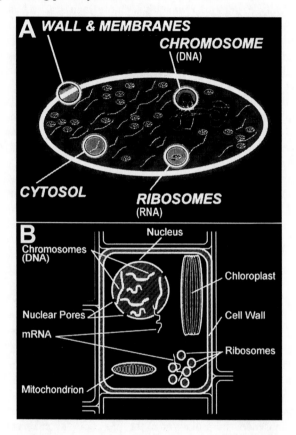

Fig. 1-10 Comparison of a prokaryote (bacterium) cell **(A)** and a eukaryote (plant) cell **(B)**. Though both cell-types are DNA-encoded with similar instructions to carry out similar basic functions and both contain enzyme-producing ribosomes, the more advanced eukaryotic cell houses various cellular processes in membrane-enclosed organelles (nuclei, chloroplasts and mitochondria).

In contrast with the cells of eukaryotic animals and plants, microbial prokaryotic cells do not contain membrane-enclosed nuclei, mitochondria or chloroplasts, their gene-carrying chromosomes and enzyme-producing ribosomes being scattered throughout the cytoplasm, the watery "cell sap."

There are other differences as well, important here because they can be detected in the fossil record. The cells of eukaryotes are relatively large, typically 5 μm (micrometers) to more than 100 μm in diameter, whereas those of prokaryotes are almost always small, only 0.1 μm to less than 5 μm across. Eukaryotes are mostly many-celled, while prokaryotes are generally single-celled. And virtually all eukaryotes multiply by sexual reproduction, combining the genes from two different parents, whereas prokaryotes are uniformly non-sexual, their cells simply dividing and the genes of the single parent passed along unaltered to the offspring. As a result, eukaryotes evolved far more rapidly, the mix of genes in their offspring adding a new source of variability to their species, while prokaryotes simply plodded along, unchanged except for rare useful changes (mutations) in their genetic instructions.

We, like all other animals, are eukaryotes. In their size, architecture, chemistry, cell organelles and virtually everything else, our cells are like theirs, their cells like ours. But humans, *Homo sapiens*, are **not exactly** like any other eukaryotes – just as frogs differ from lizards and those differ from birds and horses, dogs or cows. Each species is distinct from all the others, simply because each contains its unique species-defining set of gene-encoded instructions, in all cases derived and then modified from those of its evolutionary ancestors.

All life belongs to three great Domains

How does the two-cell-type architecture of life translate into all-encompassing categories that embrace everything that has ever lived? As you might have expected from the foregoing discussion, for years and years all were taught that there are only two Kingdoms of Life, not animals and plants but prokaryotes and eukaryotes, the primary distinction being the absence (in prokaryotes) or presence (in eukaryotes) of an identifiable cell nucleus known to house biochemicals that play a central role in an organism's life. But the nature of those chemicals – known now to be the DNA (deoxyribonucleic acid) that makes up the genes of nuclei-encased chromosomes – and what they did, how they worked, were unsolved mysteries. In 1953, the picture changed, markedly. Based in part on critical X-ray crystallographic analyses of DNA presented by the English chemist Rosalind E. Franklin (1920-1958) at a university department seminar and then used – without her knowledge or permission and even without acknowledging the source of the data – by American biologist James D. Watson (born in 1928) and English biochemists Francis H.C. Crick (1916-2004) and Maurice H.F. Wilkins (1916-2004) to establish the two-stranded double helix structure of DNA, an epoch-making discovery that unlocked the genetic code. To honor this major accomplishment, Watson, Crick and Wilkins (**Fig. 1-11A-C**) – but not Franklin (**Fig. 1-11D**) – shared the 1962 Nobel Prize in Physiology or Medicine. Although Watson is reputed to have suggested that it would have been appropriate for Franklin (along with Wilkins) to have been awarded a Nobel Prize in Chemistry, Franklin had died four years earlier and the Nobel Committee generally did not present posthumous awards.

Fig. 1-11 Co-contributors to the discovery of the structure of DNA, **(A)** James D. Watson, **(B)** Francis H.C. Crick, **(C)** Maurice H.F. Wilkins and **(D)** Rosalind E. Franklin.

Up to this time, the prokaryote-eukaryote distinction had been based on optical microscopy of the contents of cells. Then, in the 1960's, the transmission electron microscope (TEM) entered the scene, a new instrument that enabled high-magnification imaging of cellular inners. The new data provided by TEM prompted readjustment of some earlier classifications – oxygen-producing photosynthetic microbes, for example, long-classed together with eukaryotic seaweeds as "blue-green algae," were shown to lack a cell nucleus and were thus reassigned to the prokaryotes, henceforth resulting in them to be referred to as "cyanobacteria" to more accurately reflect their proper evolutionary relations.

Although such reassignments were few and far between – in retrospect, seemingly minor and unimportant – they nevertheless highlighted a glaring lack of in-depth knowledge about the nitty-gritty of the biological world. This, in turn, prompted Carl Woese (1928-2012, University of Illinois) to tackle the problem, investigating not just cyanobacteria but all the lineages of life and using the chemistry of species-distinctive genetically programmed rRNA (the ribonucleic acids of protein-manufacturing ribosomes) rather than the morphological shapes and structures revealed by TEM. What he discovered is that yes, nucleated eukaryotes fit together as a coherent whole, but that no, earlier evolved prokaryotes do not mesh together into one great group. Instead, he placed minute microbial prokaryotes into two similar but distinct tribes (Super-Kingdom-like "Domains"), the Bacteria, as they had been before, and a second major tribe that he dubbed the Archaea (from the ancient Greek word *archaios* meaning "ancient, primitive"). Interestingly, Carl later owned-up to me that he much regretted his newly coined moniker, calling it *"the biggest mistake of* [his] *career"* and an error made only because he was unaware that the earliest two-billion years of Earth history were already known as the Archean, the use of the two terms – one for primitive organisms the other for ancient rocks – causing confusion. In any case, Woese's name for the newly described group, which he and microbiologist George Fox (University of Houston) proposed in 1977 was set in place in 1990 by Woese and his colleagues, a sea-change ultimately giving rise to the now widely adopted Three-Domain classification of all living system – Archaea, Bacteria, and Eucarya (**Fig. 1-12**).

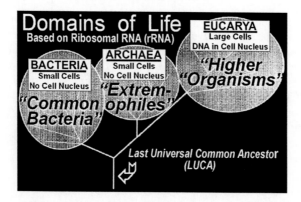

Fig. 1-12 The three great Domains of life on Earth – Bacteria, Archaea and Eucarya – a classification formalized by Carl Woese and his colleagues in 1990.

For your amusement, I here relate a brief personal anecdote. In 1985, Carl invited me to give the University of Illinois' annual DuPont Lecture. After the lecture, we ventured over to a local restaurant. There, as two of us were unwinding in the bar, all were told that a tornado warning had been announced and that we were to ride-out the storm in their basement refuge. After the others departed, Carl darted behind the bar and grabbed a gallon-jug of red wine, well more than sufficient to keep the two of us going for the following three hours of our incarceration. At about four o'clock in the morning – by then, Carl knowing about my earlier success as a trumpet-player – he took me home to listen to his new "Hi-Fi." He turned it up loud – **really loud** – so that I could enjoy a then-famous small jazz band, *The Dukes of Dixieland*. After awhile, and surprisingly to me, an enrobed woman (Gabriella, Carl's wife) emerged above and charged down the stairs: *"Carl, get your ass up here ...now!"* HA! I departed … immediately.

Now, back to the "real stuff." Woese had to demonstrate that his new Archaea Domain of microbial life differed, importantly, from the long-known "run-of-the-mill" Bacteria Domain. And this he did, showing that that the cell walls of the two groups are composed of decidedly differing chemicals, the walls of Bacteria made of peptidoglycan (known also as murein), a Bacteria-defining component not present in the walls of Archaea.

Moreover, and even more decisive, most members of the two Domains live in greatly differing habitats. Almost all Bacteria are "mesophiles," organisms that grow best at moderate temperature, neither too hot nor too cold, with an optimum growth-range from 68° to 113°F

(20° to 45°C). In contrast, a great number of Archaea are "extremeophiles," microbes that prefer what we regard as extreme conditions of temperature, pressure, acidity, alkalinity and the like … though to the Archaea, such environments are perfectly normal. Thus, Archaea are particularly abundant in hot springs and deep-sea vents, having been found more than 6 miles (10 km) deep in the ocean where the water pressure is more than 16,000 pounds per square inch (1,125 kg/cm^2); in conditions ranging from extremely acidic (pH zero) to highly alkaline (pH 12.8); in hydrothermal submarine vents at 252°F (122°C) to frozen sea water at -4°F (−20°C); and, in tiny fissures in rocks of the Earth's crust at depths of more than 4 miles (6.5 km).

The habitation by Archaea of such seemingly uncommon settings explains why they remained unknown for such a long time. But to scientists who wanted to study such microbes, this also presented huge difficulties, especially given that Archaea were only known to be abundant at all-but-inaccessible deep-sea thermal vents. This problem was solved at the University of Regensburg, in Bavaria, southern Germany. Because of the local brewery industry (think of nearby Munich's annual Oktoberfest), microbiologists at the university were able to obtain large beer-making fermenters in which they expertly grew and harvested Archaeal microbes obtained from the sampling of deep-sea vents. These vats permitted them to mimic the microbes' original setting – their extreme conditions of temperature, pressure, acidity and the like. The microbes thrived, samples of which the cooperative, generous scientists then doled out to workers worldwide. Although such special growth conditions are no longer necessary, Archaeal microbes being known today in numerous far more accessible environments than they were in the 1990's, this pioneering work allowed Woese's concepts to be repeatedly confirmed – Archaea and Bacteria do, indeed, decisively differ.

Conclusion of the Prelude

Given the foregoing overview of geological time, the workings of evolution, and the types of life on Earth, we are now poised to consider the *Blooming of Life*, the remarkable interrelated evolutionary history of the plants and animals that surround us. As you will see, these truly mind-boggling events, spanning more than 550 million years of Earth history, are explicable, understandable but often surprising. And from our perspective they *do* matter. After all, they are the "stage-setters," the long necessary evolutionary prelude to the rise of our species of life, *Homo sapiens*.

CHAPTER 2

PLANTS CONQUER THE LAND

Plants truly matter!

We all know plants – at least we "sort of know them" – they're all around us! If you were like me when I was youngster, you came to know them a whole lot better than you would have wished. After school, I wanted to head off with my pals to play basketball, football, baseball or to visit my girlfriends, but I had to do my chores first. Complete my homework and then, from time to time, mow the lawn, clean out the leaf-filled gutters, rake the leaves, and pull weeds (the dandelions with their deep taproots were the worst). Oh yes, I came to know plants!

Despite this all-too "up close and personal" introduction to the plant world, I now know that everyone of us is in fact wholly dependent on plants for our very existence. How could that be? The first reason is obvious: We eat plants – the vegans among us only plants and plant products, with the meat-eaters also feasting on scrambled or sunny-side-up chicken eggs, beef or buffalo burgers, juicy pork, and fish from sardines to salmon to tuna. But, of course, the chickens, cows, buffaloes, pigs and fish – in fact all types of animal-life – feed on plants (including algae) and plant grains (corn, wheat, sorghum and the like). In that sense, you can think of animal meat as also being a (second-generation) "plant product."

Nevertheless, and regardless of our dependence on plants for food, most of us continue to take them for granted … they're here, they're there, they don't bother us, they don't matter. That, however, is way off base. Plants matter to our well-being in a host of other ways as well. Foremost among these is their ability to produce oxygen by photosynthesis, without which we could not exist. And we also use plants and plant-products in many other life-improving ways, for housing (wood), fuel (coal), fibers (clothes, paper, rope), and beauty (landscaping, Valentine's Day bouquets), to list just a few. Moreover – and though it is little known and not widely appreciated – plants provide us medicines, like the drug digitalis (obtained from the dried leaves of the common foxglove, the herbaceous shrub *Digitalis purpurea*), a prime remedy for treating heart disease; and the drug curare, extracted from a South American woody vines (*Strychnos*), a highly effective muscle relaxant used by the indigenous Amazonians to hunt birds. Further, and particularly important, plant-generated oxygen is the source of the planet's stratospheric ozone layer that shields all Earth-life from the harmful effects of the Sun's ultra-violet rays.

Yes, oh yes, plants truly *do* matter!

The plant-animal difference

Plants, like animals, are nuclei-containing eukaryotes. All plant cells also contain chloroplasts, the intracellular organelles housing their photosynthetic apparatus, as well as mitochondria, minute intracellular energy factories (**Fig. 1-10B**). Thus, plants both photosynthesize and breathe, photosynthetically producing their life-sustaining food, glucose sugar (CO_2 + H_2O + sunlight → "**CH$_2$O**" [glucose] + O_2) and then using this self-manufactured glucose to grow and thrive by reversing the process to produce energy-storing ATP (adenosine triphosphate) via "aerobic respiration" (O_2 + "CH_2O" → H_2O + CO_2 + **ATP** [energy}. In other words, plants can both make their own food via photosynthesis and "burn" that food (combining it with oxygen) to produce energy by breathing. In contrast, animals, which lack chloroplasts, can only breathe (like plants, carrying out aerobic respiration in their mitochondria and producing energy-rich ATP).

Though most of us imagine that animals are primary in our world – after all, we too are members of the Animal Kingdom – plants are actually the more complicated biochemically, more accomplished in their fundamental life-sustaining intracellular tricks than are animals. And though a plant-animal hybrid would be interesting and less expensive for farmers to feed and maintain … for example a chlorophyllous green cow, making its own food during the day and grazing only at night … such mix-and-matchers exist only among single-celled protozoans such as *Euglena* and its close kin. In short, although animals rely on plants, the plant-world could get along just fine without us.

In the mid- to late-1800s when, thanks to Darwin's newly unveiled theory of evolution, some scholars began to wonder about how life began the biochemical differences between plants and animals and their evolutionary roots were little understood. Indeed, Darwin, focusing on the evolution of life rather than its origin went so far as to pooh-pooh even worrying about life's beginnings, in 1887 writing to his friend Joseph Hooker that "*It is mere rubbish thinking at present of the origin of life;*

one might as well think of the origin of matter." What **was** known is that animals (heterotrophs, literally "feeders on others") eat plants (autotrophs, "self-feeders"), an obvious fact that led all to assume that plant-like autotrophs must have been the first forms of life, animal-like heterotrophic eaters coming later, 'cause otherwise the primordial animals would have had nothing to eat! Not only did this seem obvious, but it was a notion that also fit with the prevailing wisdom that animals such as us are undoubtedly the more advanced of the two groups.

In 1924, this prevailing view was challenged by a freshly minted PhD student, Russian biochemist Aleksandr Ivanovich Oparin (1894-1980). Based on simple plant-growth experiments he had carried out as a high school student and spurred by the teachings of Moscow State University botanist Kliment Arkadievich Timiryazev (1843-1920), Oparin realized that plants are metabolically more complex, animals simpler. Based on this insight and Darwin's theory of evolution, Oparin's newly announced notion postulated that life began with simple animal-like microbial heterotrophs, eaters that fed on a primordial organic-rich soup, and that more complex plant-like microbial autotrophs followed later.

Here is Oparin's story of how he came up with this new notion, told to my students and me when Oparin visited my lab at UCLA for a two-month stay in 1976. As Oparin recounted to us, in 1916, as a graduating high school student in Uglitch – a small town on the Volga River northeast of Moscow – he, like many of his classmates, took the national scholastic aptitude test in preparation to move on to college. After he and one other student were selected to attend Moscow State University, Russia's finest, their science teacher arranged to take them to Moscow to have a look at the campus. They missed the morning train from Uglitch and arrived in mid-afternoon, allowing time to attend only one class session. Oparin elected to attend Timiryazev's botany lecture, primarily because he had previously read two of Timiryazev's textbooks and from them had become interested in Darwinian evolution.

As fate would have it, Timiryazev, nearing retirement, spent the hour explaining how he had become a convinced evolutionist. Born into a well-educated family, upon Timiryazev's receipt of a doctoral degree his parents provided funds to permit him to visit Darwin at his home ("Down House") in Kent, southeast of London. Upon his arrival, he discovered that Darwin was ill, as he often was in his later life, so Timiryazev journeyed down Luxted Road and rented a room at the Queen's Head Pub, since the 16th century the local village "watering hole" (which stands to this day). Each morning he returned to Down House, sat patiently on the stoop at the front door, and awaited an audience with Darwin. After several days,

Darwin appeared and the two of them walked the famous Sandwalk, the mile-long pebbled track at the rear of the property where Darwin explained his Theory of Evolution. From that beginning, Timiryazev went on to become Russia's leading (and perhaps sole) outspoken proponent of biological evolution.

As Oparin phrased it, *"Darwin revealed the evolution of animals ... Timiryazev showed the evolution of plants ... but no one had written the first chapter of the book, how life originated."* With this impetus, Oparin began his quest to provide that missing opening chapter. By 1918, only two years later, he had completed his write-up which he submitted for publication. But beginning in the autumn of the previous year, Russia had been in turmoil, beset by the 1917 Bolshevik Revolution and the overthrow of Czar Nicholas II. According to Oparin, though the Czar had been dethroned his all-powerful censors were still in place. The censors were appalled by Oparin's ideas – which meshed not at all with the teachings of the Russian Orthodox Church – and, thus, his manuscript was summarily rejected. Over the following five years he completed a much-expanded rewrite, a years-long hiatus that he claimed was *"the best thing that could have happened"* because it gave him time to bolster his heterotroph-first, primordial soup hypothesis. In 1924, his revised opus was published, a volume he affectionately referred to as his *"little pamphlet."*

Like Darwin before him, Oparin's ideas flew in the face of accepted theological wisdom. Moreover, and of appreciably greater impact on non-Soviet scholars – given the fear raised in the West by the Bolshevik Revolution – Oparin's scenario of a simple-to-more-complex natural evolutionary progression fit well with Marxist-Leninist Communism, a doctrine devoid of a need for God's direction or intervention. Indeed, it was not until Oparin's concept was later confirmed by the 1953 break-though non-biological laboratory synthesis of life-like amino acids by University of Chicago doctoral student Stanley Lloyd Miller (1930-2007) that many in the Western World came to accept Oparin's concept. Perhaps most importantly, Miller's abiotic synthesis used the same mix of primordial atmospheric ingredients as that postulated by Oparin (methane gas, CH_4; ammonia, NH_3; water vapor, H_2O; and hydrogen gas, H_2). Now, Oparin's thesis, life beginning with simple heterotrophic microbes and advancing to more complex "plant-like" autotrophic microorganisms, is accepted worldwide, its basics, if not all its nuances, confirmed. In other words (microbial) animal-like life came first, more advanced plant-like microbes later.

As this snippet about the history of science well illustrates, given time – ideologies aside – careful studies produce firm facts. Facts are facts, and facts always win. That, thankfully, is the way of science.

From life's beginnings billions of years ago, by about 450 Ma ago, during the mid-Silurian Geological Period, evolution gave rise to the land flora on which we all depend. This then raises a series of questions, not least among them is how do plants accomplish their role, how do they manage the complicated tasks they carry-out each day?

Origin of land plants

Even though plants do biochemical tricks that animals cannot, their basic workings are not difficult to understand. The first stages of their colonization of the land surface provide a good way to think about the problem, so let's start at the beginning. Where did land plants come from? What developments were needed for them to invade the land? How and why did these changes matter?

Like algae, "seaweeds" and "pond-scum," their distant evolutionary precursors, plants are photosynthesizing autotrophs. And the earlier transition from marine algae to brackish- and fresh-water algae was not difficult, largely a matter of adjusting to a less-salty setting. In contrast, the step from a fresh-water environment to that of the land surface required a lot of changes.

Water, of course, is needed by all forms of life and especially by plants without which they could not carry out photosynthesis (CO_2 + H_2O [**water**] + sunlight $\rightarrow$ "CH_2O" + O_2). For algae this is no problem – they are surrounded by water, which they can easily absorb. But for land plants, the source of water is the ground below whereas their photosynthesis takes place above, in their stems and leaves. So, to make the transition from water to land, plants had to develop means to absorb water from the ground and then pipe it up to their chlorophyll-bearing organs where photosynthesis occurs. Today, the plants we know absorb groundwater by their roots, fairly fancy appendages that the first land plants did not have. Instead, they used their underground stems ("rhizomes," similar to the underground stems of the bamboo in your neighbor's garden that uses them to sneak into your home garden) and then piped this water up to the rest of the plant through a newly evolved tissue known as "xylem." The xylem tubes, stout, vertical, straight and hollow like soda straws, form the central core of plant stems. They are present in all land plants – in trees they make up the tissue we call wood. Because their tubular form is maintained and strengthened by spirals or ring-like bands of a secondarily plastered-on particularly sturdy organic compound, lignin – the addition of which making them what are called "tracheids" – the water-conducting tissue they comprise gives flexibility and strength to the plant stem. In essence, these lignin-reinforced exceedingly narrow hollow xylem tubes are rather like the long steel rods of steel-reinforced cement towers **(Fig. 2-1A)** in which the rods, like the xylem tubes, provide a degree of flexibility – permitting the steel-reinforced towers to sway during an earthquake – and the cement plastered around the rods, like the lignin spirals and bands that encase the xylem tubes, providing strength **(Fig. 2-1B)**. (Note that you already "knew" this – without really knowing it – by what builders refer to "hard woods," those that are lignin-rich and "soft woods," those relatively lignin-deficient.) The earliest land plants were thin and short, little more than a foot or so in height, so their stem-supporting central xylem core was also tiny. walls providing flexibility (like the steel girders), the surrounding lignin bands (like the cement casing) providing strength.

So far, so good – new evolutionary inventions to absorb water and then pipe it up the plant – yet still more changes were needed. Getting water up the plant would do no good if it there simply evaporated out into thin air. To defeat this problem, early land plants evolved a waxy skin by using the water-repellent biochemical "cutin" to coat their outer surfaces, a cutinized epidermis. Though this skin retained the water, it shut out the other needed ingredient for photosynthesis, CO_2, a problem solved by the invention of small surface holes, pores known as "stomates," that open during daylight hours to promote photosynthesis and close at night, to keep the water in. Still, the life-sustaining product of photosynthesis, the sugar glucose, had to be transported from the sunlight-lit surface source back into the rest of the plant, the blacked-out center of the stem where photosynthesis cannot occur. This task was accomplished by the invention of phloem, a second set of thin-walled tubes – a bit like narrow, long flexible noodles – that in modern plants run parallel to the xylem strands to make up water- and food-conducting veins of a leaf.

Pause for a moment and ponder. Like plant leaves, we have veins too – to pipe oxygen-rich blood through our bodies, not water and photosynthesis-made glucose – but the tubular structure and flexibility of the two types of veins are basically similar and both perform similar functions. This is a good example of the similar problems encountered by plants and animals that have led to similar solutions, an important feature of the plant-animal parallel evolution that we will see time and again in the evolutionary history of the Blooming of Life.

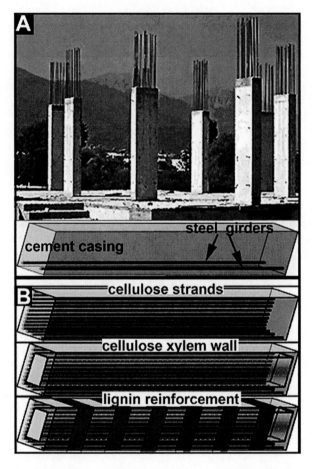

Fig. 2-1 (A) Steel reinforced cement towers under construction. **(B)** Comparison of steel-reinforced cement with a lignified xylem (tracheid) tube, the cellulose xylem

Were rhizomes, xylem, lignin, cutin, stomates and phloem – all traits that have carried through to plants today – enough to complete the transition from water to land? Almost, but not quite. The earliest land plants inhibited marshes and the margins of the nearby dry land, a necessary first step to their ultimate complete invasion of the land surface. And though they were enormously successful in this new non-competitive setting, they had to live near water because of the way they reproduced. Like fern plants today, the earliest "vascular" plants, those having xylem and phloem, reproduced by spores that matured, were spewed onto the ground and produced sperm and egg with the sperm (like human sperm) swimming to fertilize a nearby egg. To carry out this role they needed a moist surface, like that in a marshy setting. Moreover, the spores themselves were vulnerable and would have rapidly dried, collapsed and died – a sure pathway to the ultimate extinction of their parental plant stock. To offset this, the spores were coated with same waxy cutin that protected the plant stems.

Finally, then – with all of these evolutionary advances (rhizomes, xylem, lignin, epidermal cutin, stomates, phloem and cutinized spores) – plant-life made its move.

The rise of plants on land

Although land plants were not the first occupants of Earth's land surface – an accolade reserved for their forerunners (microbial photosynthetic cyanobacteria, diverse other ancient microbes, and microscopic algae and filamentous fungi) – the rise of plants on land is well-documented in the fossil record, the first big finding being that of the early Devonian (410 Ma) Rhynie Chert Flora more than a century ago. Discovered in 1912 near the village of Rhynie in Aberdeenshire, Scotland by Dr William Mackie (1856-1932), a Scottish physician well remembered for his contributions to geology, this discovery set science on its heels. The evidence was unassailable, meticulously documented by British paleobotanists Robert Kidston (1852-1924) and William Henry Lang (1874-1960) in a five-part monographic series published from 1917 to 1921 that showed this to be the oldest in situ-preserved terrestrial ecosystem known. Cellularly permineralized in fine-grained chert – like the Gunflint and Bitter Springs Precambrian microbiotas noted in the Preface to this book – the flora included five distinct types of early land plants, a real milestone because it was the first time that higher plants showing such a great amount of anatomical detail were described from rocks this old. Even today, a full century later, the Rhynie Flora provides the basis of our understanding of early land plants. (As described by Kidston and Lang in the 5th volume in their series, the flora also includes bacteria, cyanobacteria and microscopic fungi, their mention added almost as an afterthought.)

In comparison with plant-life today, the higher plants of the Rhynie Flora were simple, "primitive." Thus, the most abundant and least advanced member of the assemblage, in 1921 named by Kidston and Lang *Rhynia major,* lacked true roots, had a lignin-lacking and thus only partly evolved xylem core, and was leafless, its naked green chlorophyll-bearing stems little more than 10 inches (25 cm) high, revealing a very early stage in the rise of plants on land **(Fig. 2-2)**. As you might have imagined – given that *Rhynia major,* like all other early evolved plants reproduced by spores that needed a moist substrate for their sperm to swim to fertilize the plant's egg cells – the Rhynie deposit is a preserved marsh, the plants growing in an intermittent wet-dry setting. Their water-conducting xylem core **(Fig. 2-3A)**, well documented by Kidston and Lang, was small, needed to support only their short height, and preserved in near-life condition like all of the other plants of the assemblage their stems are short, thin, erect, and exceedingly close-packed **(Fig. 2-3B)** because they lacked spread-out leaves that would have required additional space. With the exception of *Rhynia major*, the xylem cores of the Rhynia plants were composed of true tracheids – that is, they were "vascular plants" – their water-conducing tubes reinforced by spirals and or rings of lignin. Such water-conducting tracheids are present in all modern plants, in trees providing the strength and flexibility of their woody trunks.

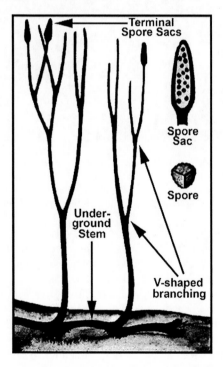

Fig. 2-2 *Aglaophyton* ("*Rhynia major*") in growth position, illustrating its stem-terminating spore sacs, V-shaped branching pattern, and underground stem (rhizome).

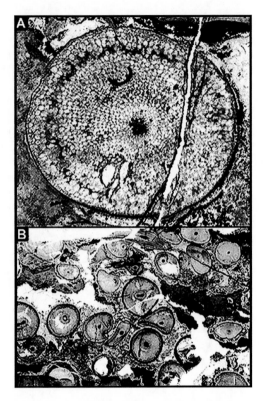

Fig. 2-3 (A) Cross-section of an *Aglaophyton* ("*Rhynia major*") stem showing its tissues and the small size of its central water-conducting thick-walled xylem core. **(B)** Overview of the cross-sections of multiple specimens of this rhyniophyte in growth position, showing their close packing in the Rhynie Flora-hosting deposit.

Given the primitive, only partly evolved (non-lignified) water-conducting xylem core of *Rhynia major,* a prime factor indicating its midway position between algal-derived bryophytes (such as mosses and liverworts) and modern land plants, tracheophytes, in 1968 British paleobotanist Dianne Edwards (Cardiff University, Wales; **Fig. 2-4**) – an internationally renowned paleobotanist and much honored CBE (Commander of the Most Excellent Order of the British Empire) and FRS (Fellow of the Royal Society) – reassigned *Rhynia major* to the new genus *Aglaophyton* (from the Greek *aglaós,* "splendid, beautiful, bright" plus *phyton*, "plant" – "beautiful plant"). Nevertheless, like almost all components of the flora, *Aglaophyton* ("*Rhynia major*") has naked stems that branch in a simple algal-derived V-shaped (dichotomizing) form, and features as well the newly evolved innovations of stem-terminating spindle-shaped spore-sacs and a water-conducting central core. Moreover, although the other plants of the flora have a more advanced lignin-fortified water-conducting xylem core, all of the plants, *Aglaophyton* included, have a zone of food-conducting thin-walled phloem-like tissue encircling their thick-walled water-conducting core, a cutinized stem-enclosing epidermis, and spindle- or kidney-shaped spore sacs (sporangia). Similarly, the aerial axes of all Rhynie chert plants consist primarily of a thick layer of cortical tissue (parenchyma) that surrounds the centrally situated xylem water-conducting strand that in virtually all the plants constitutes only a small portion of the total stem.

Fig. 2-4 World-renowned British paleobotanist Dianne Edwards.

Notably, almost all of the components of the Rhynie Flora are entirely leafless, naked, the single exception being the tallest and most advanced, *Asteroxylon mackiei*, named by Kidston and Lang for its star-shaped xylem core **(Fig. 2-5)** and to honor the discoverer of the deposit, William Mackie. This plant, among the earliest known to exhibit true leaves (and informally referred to as "ole star-wood" because of its

geological age and the radiating shape of its xylem core) has numerous
spores sacs attached to the sides, not the tips, of its stems, an innovation
resulting in an increase in the number of spore sacs and thus spores
produced and dispersed that increased the survival of its species.

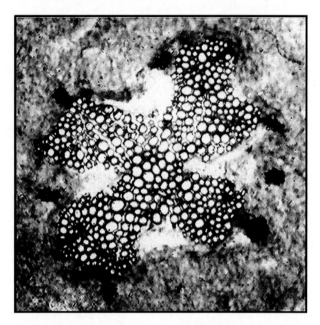

Fig. 2-5 In comparison with the cylindrical small central water-conducting core of
Aglaophyton (**Fig. 2.3A**), the lignified xylem core of the most advanced plant of
the Rhynie Flora, *Asteroxylon,* is star-shaped, its lobate peripheries extending
toward the sides of the stem.

Considered as a whole – with the exception of *Asteroxylon* –the
plants of the Rhynie deposit are lumped together as members of the
botanical group Rhyniophytina, more informally "rhyniophytes," *Rhynia*-
like plants.

Despite the marvelous insights into the early evolution of land
plants provided by the Rhynie Flora, they are not the earliest vascular land
plants now known. Instead, that distinction goes to the even earlier-
evolved genus *Cooksonia* (**Fig. 2-6**) discovered in 1937 by W.H. Lang in
mid-Silurian (440 Ma) rocks and named by him to honor the pioneering
Australian paleobotanist Isabel Clifton Cookson (1893-1973; **Fig. 2-7**).
From this early beginning, *Cooksonia* remained a principle component of
the flora until the end of the early Devonian (390 Ma), a notable 50-

million-year-long stint. Like *Aglaophyton, Cooksonia* fossils have lignin-lacking only partly evolved water-conducting xylem tissue and, like *Aglaophyton* ("*Rhynia major*") it is thus a transitional form between the primitive non-vascular bryophytes and true vascular (trachyophyte) plants.

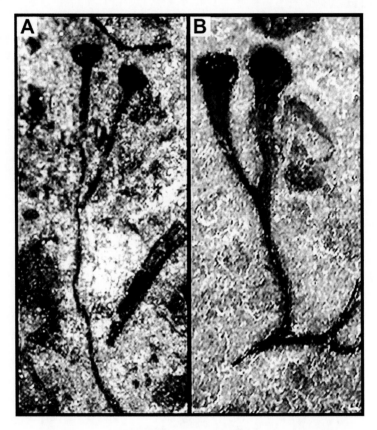

Fig. 2-6 (A, B) Two specimens of *Cooksonia*, the oldest (mid-Silurian) land plant known, showing its V-shaped branches, globose terminal spore sacs, and **(B)** its underground water-imbibing root-like stem (rhizome).

After the appearance of such naked-stemmed plants as *Cooksonia* and *Aglaophyton*, the next major advance in land plant evolution was the development of leaves such as those of the Rhynie plant *Asteroxylon*. Why do leaves matter? After all, the chlorophyll-packed green stems of plants like *Cooksonia* and *Aglaophyton* were perfectly adept at soaking up sunlight. The critical issue is that plants, like all organisms, compete with those around them – for plants, a competition for sunlight – and by the time of the Rhynie Flora, the tallest among them, *Asteroxylon*, was beginning to win the battle. Still, leaves are relatively complicated, even small simple ones like those of "ole star-wood." Where did they come from?

Fig. 2-7 Australian paleobotanist Isabel Clifton Cookson, after whom *Cooksonia* was named.

Here, the gap was bridged by another early-evolved plant group, the Zosterophyllophytina, the group name derived from its best-known genus, *Zosterophyllum* (from Geek *zōstēr*, a form of girdle or belt worn by men in ancient *Greece, and phyllon*, "leaf" – "belt leaf"). Zosterophyllophytes, more simply "zosterophylls," had clumped or isolated spore sacs attached to the sides of their stems **(Fig. 2-8)** – many spore sacs per stem, not just one at the tip of the stem as in their rhyniophyte ancestors – the increased number of spore sacs resulting in a greater number of spores produced and, thus, better odds in the competition to reproduce and survive. As a group, the zosterophylls range from having completely smooth stems to those peppered with randomly arrayed bumps and small spines (known also as "enations"), protuberances that increased the total stem surface-area where photosynthesis could occur. This increased area for photosynthesis, in turn, promoted an increase of plant girth, a thicker stem to support an increase of height and thus better means to compete with the other nearby plants. So successful were these spiny bumps and protuberances that they served as a prelude to the later development of true small leaves, as in *Asteroxylon* **(Fig. 2-9)**.

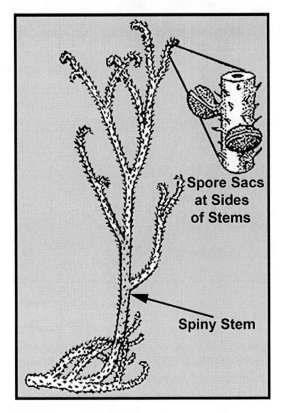

Fig. 2-8 Representative member of the Zosterophyllophytina, the zosterophyll *Sawdonia,* showing its spiny stem and side-attached spore sacs.

**Spore Sacs
at Sides
of Stem**

**Spiral Leaves
(Microphylls)**

Fig. 2-9 Earliest known lycophyte (*Asteroxylon*) showing its leafy (microphyllous) stem and side-attached spore sacs

In summary, rhyniophytes (naked stems and terminal spore sacs) gave rise to zosterophylls (naked or spiny stems and lateral spore sacs) from which evolved *Asteroxylon* and its close relatives, members of the Lycophyta (from the Greek *lýkos*, "wolf" plus *phytón*, "plant"), "lycophytes," the earliest group of plants having true leaves **(Fig. 2-10)**.

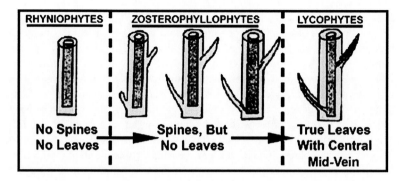

Fig. 2-10 Evolution of microphylls, the earliest true leaves, from the naked stems of rhyniophytes, to the naked, bumpy and spiny stems of zosterophylls, to the microphyllous leaves of lycophytes, each having a leaf-defining central mid-vein.

Spore plants rise to dominance

The chief difference between the true leaves of lycophytes and the short spines of zosterophylls was not just the leaves' larger size but, of greater importance, the presence of a water-conducting xylem and food-conducting phloem central strand, a leaf-defining 'vein' that is connected to the central water- and food-conducting core of the plant. The earlier evolved spines lacked this feature without which they could only be short and stubby, there being no way to pipe water from the ground to the margins of the spines and no way to transport the glucose there produced by photosynthesis back to the remainder of the plant. Moreover, because the lignified xylem in leaf veins is strong and robust, it provides support to a leaf that is lacking in such spines, even if the leaves have only a single such mid-vein running down their centers like those in *Asteroxylon* and its lycophyte close-relatives.

To botanists, simple early-evolved single-veined leaves like those of lycophytes are known as microphylls (literally, "small leaves"), their invention being a huge step forward in plant evolution. Plants compete to reproduce and to increase their height to better soak-up sunlight. Stems side-loaded with numerous spore sacs solved the first of these problems and now, with the development of true leaves, microphylls, the second problem had been put to rest – by simply packing their stems with leaves, a leafy plant and its descendants would win the competition. Sounds good! But it has a flaw, namely that if the leaves are tightly packed in a haphazard irregularly arrayed pattern, like the spines of zosterophylls, some leaves below will be shaded and thus rendered useless by those above – and the more closely packed the leaves, the more that will be shaded below. Thus, for even a stem packed with lots of leaves, many of the down-stem leaves would not function because they could not "see the Sun."

Here, again, evolution answered the problem. On the pathway from zosterophylls to lycophytes like *Asteroxylon* "ole star-wood"), plants developed the genetic processes needed to arrange their light-catching leaves into a pattern that worked far better, organizing them in a stem-encircling spiral so that the leaf above would not shade a leaf directly below. Later-evolved lycophytes carried this problem-solving and life-sustaining principle through to all of their organs – spirally arranging not only the leaves on their stems, but the leaves on their branches, their spore sacs on the upper surfaces of these leaves (leading to spirally close-packed spore cones), and even to their roots, their protruding water-imbibing rootlets arranged in a similar spiral pattern.

Given this final great advance, lycophytes took over the swamp, the best-known example being *Lepidodendron*, the giant scale-tree of the Carboniferous, about 320 Ma ago living in the less-salty margins of vast inland seas. Thus, scale-trees and other closely related microphyllous plants were major components of the mid-America "Coal Swamp Flora" for which the Carboniferous Geological Period is named. Their carbon-rich fossilized remains, in turn, provide the basis the coal-mining industry that parallels the present-day upper reaches of the Mississippi River and its northern tributaries, the river valleys being remnants of large inland seas that inundated the central continent. In terms of the geological past, this Carboniferous flooding was by no means odd. In fact, over Phanerozoic time, as worldwide sea levels have episodically risen and fallen, the central United Stares has repeatedly been flooded by just such vast inland seas. The Carboniferous Coal Swamps were essentially "normal," the present dry habitable mid-continent being something of an anomaly (as is evidenced by the Global Warming-induced ever-increasing lowland flooding of the Mississippi delta near New Orleans).

Young *Lepidodendron* trees had spirally arranged leaves encircling their stems **(Fig. 2-11A)** that ultimately sloughed off (much like the yearly shedding of leaves of the many leaf-litter producing "deciduous" trees, such as the maple, around us today). Thus, as L*epidodendron* matured, these sloughed-off leaves left a distinct tell-tale pattern of spiral leaf bases that make their fossil stems easy to identify. Mature trees were tall, like giant telephone poles, as much as 150 feet (45 m) in height and typically had large, up to one-and-a-half-foot- (0.45 m-) long heavy spore cones hanging pendulously from their branching leafy crowns **(Fig. 2-**

11B). The shallow root system of *Lepidodendron* is an "organ genus" known also as *Stigmaria* (from the Latin *stigma,* "mark" plus the suffix – *aria*) that was named separately from *Lepidodendron* because its fossils are commonly found no longer connected to their rotted-away parent stems. In mature plants the root system was remarkably expansive, 40- to 80-feet (12- to 24-m) across **(Fig. 2-12A),** splayed just below the muddy marshy surface. Each root was equipped with a large set of spirally arranged closely spaced rootlets – which, like the leaves, sloughed-off to leave characteristic spiral rootlet scars **(Fig. 2-12B)**. The combination of great plant height and huge splayed-out root system immediately tells you about their environment, their distinctive shallow root system needed to spread the weight of the tall telephone-pole-like trees in a soft, swampy, marsh-like setting.

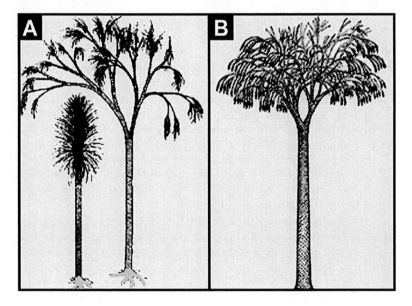

Fig. 2-11 Carboniferous scale-tree *Lepidodendron.* **(A)** Juvenile trees; **(B)** mature tree with pendant spore cones.

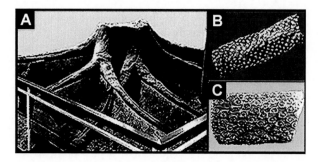

Fig. 2-12 Specimens of fossil *Stigmaria*. **(A)** A mature root system (Manchester Museum, University of Manchester, England). **(B, C)** *Stigmaria* roots showing their characteristic spirally arranged rootlet scars.

One other type of microphyllous plant was especially prominent in the Coal Swamp Flora, members of the Sphenophytina (from Greek *sphēnoeidēs*, "wedge-shaped" plus *phytón*, "plant"), "sphenopsids," forerunners of the modern genus *Equisetum* (from Latin *equus*, "horse" plus *saeta*, "bristle, hair") that are commonly known as "horsetail plants." The Carboniferous members of the group, up to 30 feet (9 m) or so in height **(Fig. 2-13A),** are members of the genus *Calamites* (from Latin *calamus*, "reed, cane"). Stems of *Calamites* were hollow and vertically ridged both outside and on their interior surfaces. When a plant died, this hollow cavity, the pith of the stem, could become filled with washed-in fine sediment, and when the surrounding plant wall rotted away this left a telltale cast of the interior, its fossilized "pith cast" and the most common way that *Calamites* is preserved **(Fig. 2-13B).**

Like lycophytes, sphenophytes had spore cones and single mid-veined microphylls, but instead of being organized into a spiral pattern like lycophytes the sphenophyte stems were segmented into a series of circular branch- and leaf-producing rings, called nodes, between which were longer inter-nodal regions. This node-internode organization is evident both in fossil *Calamites* **(Fig. 2-13)** and in the stems of the still-living horsetail plant, modern *Equisetum,* the stems of which, like those of *Calamites*, are vertically ridged and hollow. Leaves and branches emerged from these nodes – but not from the intermodal areas – the branches having the same node-internode organization as the stems and their microphyllous leaves being arranged in disc-shaped rings. Thus, whereas lycophytes were "spiral plants," their external plant parts arranged in regular helical swirls, sphenophytes were "disc-plants," their discoidal arrangement being a simple condensed evolutionary derivative of their spirally organized ancestors.

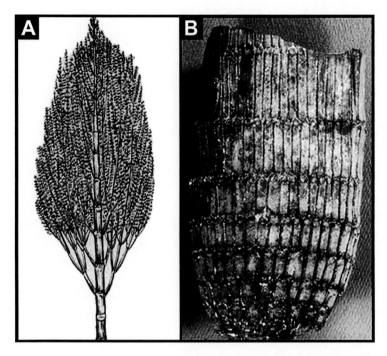

Fig. 2-13 *Calamites*, the dominant Carboniferous sphenophyte. **(A)** Mature plant, about 30 feet in height. **(B)** Fossil *Calamites* pith cast, the most common mode of preservation of this sphenophyte, showing its vertically ridged stem and node-internode organization.

These early-evolved lineages – the leafless rhyniophytes and zosterophyllophytes and the microphyllous lycophytes and sphenophytes – set the stage for the development of the plant life around us today. Yet, except for a very few living species they no longer exist. Where did they go, what happened to them? The answer is that virtually all perished as evolution proceeded and later-evolved plants out-competed them – the better-fit, better-adapted winning, the less fit supplanted by their more successful descendants. Thus, there are no living rhyniophytes or zosterophylls, only two remaining genera of lycophytes (the "club-mosses" *Lycopodium* and *Selaginella*) and only one living genus of sphenopsids (*Equisetum*). As we will see time and again, that is typical of all of plant (and animal) evolution as the Earth has bloomed with life. As the old saying has it, *"one day you can be the 'cock of the walk,' the next day a feather duster."*

CHAPTER 3

FROM SPORE PLANTS TO SEEDS AND FLOWERS

Spiny microphylls to large megaphylls

The earliest land flora of naked-stemmed and spiny or microphyllous shrubs differs vastly from that of today's forests and the tall leafy trees that line our streets, the plants we see about us. And except for occasional stands of landscape-beautifying rare "ornamentals" like the modern horsetail plant *Equisetum*, most of us have never seen the few surviving relicts of the beginning stages of plant life's invasion of the land, virtually all having died out long ago. Yes, we understand that they were out-competed, lost the battle and are now almost all extinct. What is it about the modern flora that caused it to win this endless competition to survive?

The result is there before us in plain sight, a flora dominated by leafy giants, but why? In other words, how and for what do plants compete? As above (**Chapter 1**), the "First Law of Biology" is to stay alive and for plants to accomplish that feat, they need only to compete with others in their immediate surroundings. But how, for what do they compete? All plants, whether primitive or advanced, have the same fundamental needs; all are photosynthetic autotrophs the basic solutions to which were solved early on as plants invaded the land. So, plants do not markedly compete for the two chemical ingredients of photosynthesis – not water, all having efficient root systems and the plumbing necessary to pipe it up from the ground through their stems, and not carbon dioxide, all being able to absorb CO_2 from the atmosphere. Given this, we are left with the third basic ingredient necessary for photosynthesis, sunlight, which, as we have seen before in the rhyniophyte-zosterophyll-lycophyte sequence can be a huge driver in plant evolution. And that, of course, is the key.

So, the prime question is how was it that some plants developed means to absorb sunlight better than their competitors. After all, the better a plant can soak up the sun, the stouter, taller and faster it can grow and that's what wins the competition with its nearby neighbors. Here, again, the answer resides in evolution, in particular the development of large leaves, "megaphylls," flat planar plate-like leaves that are far better designed to soak up sunlight than the small single-veined microphylls of lycopsids and sphenopsids.

This question then brings to the fore its two underlying components, namely what were the earliest megaphyll-bearing plants and how did they acquire this hugely important and successful trait?

The answer to the first of these sub-questions is evident from the fossil record: fern plants. Ferns and their allies, represented in the living world by some 13,000 species and known formally as members of the "Polypodiophyta" (from the ancient Greek *polu-*, the prefix poly- "many", *poús*, -pod, "foot," plus -phyta, used in taxonomy to denote a major biological category) are reported from the fossil record as early as the beginning of the Carboniferous, about 360 Ma ago.

Like rhyniophytes, zosterophylls, lycopsids and sphenopsids, ferns are spore-plants, all types of which have two distinct phases in their life cycle, a spore-producing phase (the "sporophyte generation," the part of the life cycle we refer to as the adult plant) and a gamete- (sperm- and egg-) producing phase, the "gametophyte generation." Because the spores are produced by "sex-cell division" (meiosis, known also as "reduction division"), each spore contains only half the complement of chromosomes (the haploid, "1N" number) of that in their sporophyte parent cell (the diploid, "2N" number). In ferns and other spore-plants, the spores, released from the spore sacs (sporangia) then fall to the ground where the spore cell divides by normal body-cell division (mitosis) to make a new, flat, lettuce-leaf-like independent small plant (the haploid gametophyte). After many such cell-divisions as the gametophyte grows to less than an inch or so in size, a "genetic clock" clicks on and instructs the little gametophyte plant to make sperm and egg at its opposite ends. After this, the sperm has to swim across a surface to fertilize an egg, sometimes the egg in the same gametophyte ("selfing") but more commonly that of another nearby gametophyte ("out-crossing"). Such fertilization, resulting in the combining of the haploid (1N) chromosomes of the sperm and egg, then restores the chromosome-count to the diploid (2N) number of an adult plant. The fertilized egg then divides, by normal mitotic body-cell division, to grow into a large adult diploid sporophyte plant and the cycle repeats, a process known to botanists as "alternation of generations" **(Fig. 3-1)**.

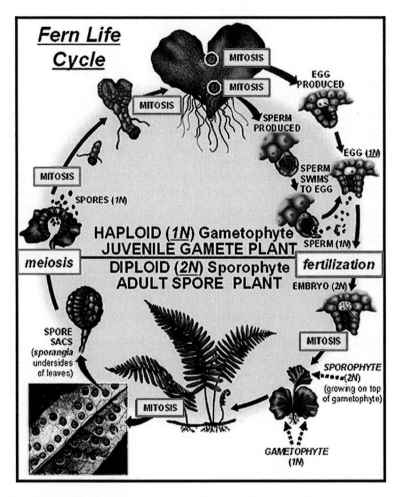

Fig. 3-1 Fern life-cycle illustrating the alternation of generations between the adult spore-producing (sporophyte) and juvenile gamete- (sperm- and egg-) producing (gametophyte) stages.

To us, acquainted with the human sperm-egg reproductive process, this spore-plant life cycle seems arcane and a bit difficult to fathom. After all, human sperm cells, with their haploid complement of chromosomes, perform only one trick, the matter of fertilizing a similarly haploid egg, whereas haploid plant spores divide, grow, and only later switch to making sperm and egg. Spore-plants have added what seems to us an extra wrinkle, the cell division of haploid spores that precedes the production of sperm and egg. However, that actually is not so odd. Seed plants do the same thing, their pollen grains, for example, being composed of four haploid cells, each playing a distinct role and the result of only two mitotic cell divisions (rather than many) of their pollen-producing parent spore. Moreover, spore-plants and seed plants were present literally hundreds of millions of years before humans – both existed long before *Homo sapiens* finally evolved – and humans and all other animals are only distantly related to the Plant Kingdom. It is thus not surprising that plants have their way of reproducing, animals have theirs.

Still, there are some similarities. Just as the sperm of a spore-plant has to swim to a haploid egg, in human reproduction the haploid sperm has to swim up the uterus to fertilize an egg (a process that in humans results in a frenzied onslaught of the egg until one of the sperm finally breaks through the egg cell's tough outer covering). And, of course, for spore-plants the "sperm must swim" requirement is the reason they inhabit the moist or marshy environments needed for their sperm to swim across a moist surface to fertilize nearby eggs. For fern plants, however, you actually already knew all this from your personal experience. In their natural environment, ferns occupy the ground-surface – for example, the understory of the Amazon rainforest – and even when they are used for landscaping around homes or on college campuses you'll find them near the sprinkler heads. Moreover, if you stop and have a look at them you'll see that their leaves are a deep shade of green (like the house-plants you might have on a table-top at your home), a darker hue of green than that of bright green grass or the leaves of street trees and an adaptation to low light-intensity. Ferns were common and abundant in the understory of the Carboniferous Coal-Swamp forests, known largely through their fossilized leaves (such as *Pecopteris*; **Fig. 3-2A**), the foliage not only of ground-hugging ferns but of taller tree-ferns as well (for example, *Psaronius*; **Fig. 3-2B**).

Thus, the earliest megaphyll-bearing plants were ferns, but how did their megaphylls arise? The leaves of modern ferns – whether narrow like many ferns or particularly broad like those of the Bird's Nest Fern – seem vastly different from the leaf of a maple tree, for example. Yet the large many-veined leaf of a maple tree, like those of all of the flowering plants we know is also referred to as a megaphyll. How did such megaphylls originate?

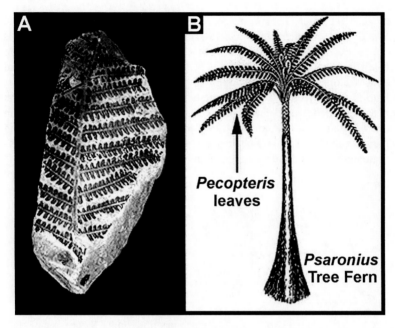

Fig. 3-2 (A) Fern leaves of the form-genus *Pecopteris* comprise the fronds of **(B)** *Psaronius*, Carboniferous tree ferns up to 30 feet (10 m) in height.

Here, again, the fossil record holds the answer. So, hark back to the rhyniophytes, the very earliest vascular plants to have inhabited the land surface. How could such primitive rhyniophytes, with their simple naked stems, V-shaped dichotomizing branch-pattern and stem-terminating spore sacs – like *Aglaophyton* **(Fig. 2-2)** and *Cooksonia* **(Fig. 2-6)** – have given rise to such a seemingly complex structure as the megaphyll leaf of a maple tree?

The story is straightforward. Initially, rhyniophytes won the land surface but they then later found themselves in competition with their descendants, spiny zosterophylls, and soon thereafter with the zosterophyll descendants as well, microphyll-bearing lycophytes **(Chapter 2)**. The rhyniophytes lost out, as did the zosterophylls, out-competed by the larger, taller, microphyllous lycophytes and their descendants, the sphenopsids. But while all this was occurring, the primitive rhyniophytes gave rise to yet another major plant group, the fern lineage, with their megaphyllous leaves.

Here's how megaphylls, the large planar leaves of the seed plants around us, initially came to be – a synopsis of what paleobotanists refer to

as the "Telome Theory" proposed in 1945-1952 by German botanist
Walter Max Zimmermann (1982-1980). According to his theory, megaphylls
– a successful innovation in ferns that was carried over in modified form
to all later-evolving seed plants – originated by changes in the terminal
ends (telomes) of the dichotomizing branch-system of rhyniophytes.
During this sequence **(Fig. 3-3)**, the telomic tips of the branch system first
became flattened and planar, rather than three-dimensional as they were in
most rhyniophytes. And second, a thin web of chlorophyll-containing
photosynthetic tissue was spread between and supported by the now-
flattened telomes. Voilà … a megaphyll!

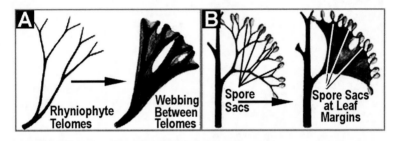

Fig. 3-3 The two principal stages of Zimmermann's Telome Theory of the origin
of megaphyll leaves, **(A)** the installation of a photosynthetic webbing between
flattened dichotomizing rhyniophyte axes (telomes) to produce a megaphyll in
which **(B)** the telomes terminated by spore sacs (sporangia) resulted in the
sporangia being situated at the leaf margin at the ends of telome-derived
dichotomizing vein systems.

From the viewpoint of a structural engineer, the telome theory
makes great sense. The goal of the plant-versus-plant evolutionary
competition is to survive, reproduce, and thus "win." To accomplish this,
the competitors battle to soak up sunlight so they can grow faster, stouter
and taller than the plants around them, a sequence well exemplified by the
naked-stemmed rhyniophyte to spiny zosterophyll to microphyll-bearing
lycophyte evolutionary sequence. But the initial skirmish-winning microphylls
of lycophytes, such as those of "ole star-wood" (*Asteroxylon*), and of
sphenopsids such as *Calamites* and *Equisetum,* are relatively small,
limited in their spread even in Carboniferous scale-trees (*Lepidodendron*)
by being physically supported by only a single mid-vein. To ultimately
win this on-going battle, a new leaf-design was needed, one that produced
more leaf-supporting veins with a layer of thin light-absorbing tissue
arrayed between them, a design that would permit this now larger planar
foliar organ to be oriented to absorb sunlight. Megaphylls, first known

from fossil ferns, fill the bill. They are both effective and lightweight, fulfilling the necessary structural engineering requirements, a lot like the lightweight umbrella you pull out in a rainstorm with its radiating struts being like the veins of a megaphyll and the rain-repellent webbing spread across them like the thin light-absorbing spread-out megaphyll leaf surface.

This is a nice story. It fits the fossil record, it makes structural engineering sense, and, seemingly, it solves the problem. But is it really true? To reassure ourselves we turn again to the fossil record. As we have seen, rhyniophytes not only have V-shaped dichotomizing branches but those stems end in terminal spore sacs. Thus, If their stems were to be flattened out, pared-down to only its water- and photosynthate-transporting xylem and phloem, and a thin web of chlorophyll-rich tissue were to be spread across them **(Fig. 3-3)**, the resulting megaphyll leave would be expected to have dichotomizing veins that end at spore sacs situated at the margins of the resulting leaf.

Indeed, that is precisely the situation shown both by early-occurring fossil ferns and by primitive ferns living today **(Fig. 3-4)**. These additional lines of evidence show that the predictions of the telome theory are borne-out, not only by the fossil record but by members of the modern flora as well.

Accumulation of such additional facts and a never-ceasing re-evaluation of an overarching interpretation are typical of the way science proceeds. It may take years, perhaps decades, as understanding advances, yet at the end of the day firm facts always win. If some new-fangled seemingly odd notion comes to the fore – like the proposal that naked plant stems gave rise to the leaves of maple trees – most scientists will initially harbor doubt (*"I never heard that before ... it must be wrong"*). Over time, the naysayers might then switch to *"OK, I sort-of agree – at least tentatively – with the concept."* Yet scientists are cautious, they always keep their guard up, never accepting a new notion without confirmatory evidence. As cumbersome as this might seem, it is exactly what is needed to get the story straight. After all, the doubting if knowledgeable experts – in this instance botanical and paleobotanical purveyors of "science by authoritative assertion" – are not **always** right. In essence, scientists thus follow the rhyming Russian proverb *doveryay, no proveryay*, "trust but verify," a useful axiom that if you embody it in your memory banks will help carry you through life. The underlying question for the origin of megaphylls was whether Zimmermann's Telome Theory could be confirmed. As the evidence has shown, it could be and has been – though the process took many years.

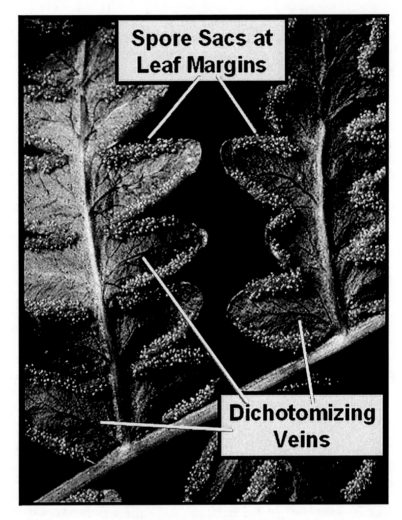

Fig. 3-4 Leaves of the primitive living fern *Pteridinium* showing its dichotomizing vein system and marginal sporangia.

Spore plants to trees with naked seeds, seed ferns provide the link

Ferns, the only group of early-evolving spore-producing plants to have continued to thrive to the present, seem far-removed from the seed plants that dominate the modern world. For example, if you have a home garden you do not visit the local gardening store to purchase a bag of spores, you are seeking a bag of seeds, whether of some handsome flowering plant or a food-providing crop-plant. What *is* a plant seed and why did their evolution matter – as it obviously must have, given that seeds are part and parcel of virtually all modern plants.

As we all know, seeds, unlike spores, are encased in a tough impermeable outer shell, the "seed coat." This protects the developing embryo within them from drying out, the protected embryo destined to become the plant's next-generation "seedling." And this, in turn, means that unlike spore-plants, seed plants do not need to inhabit moist or marshy lands to reproduce. But if that is true, how does a seed plant's haploid sperm find its way to a seed-enclosed haploid egg so that fertilization can produce a diploid seedling? To solve this problem, evolution came up with yet another innovation, wind-borne pollen grains, four-celled hay fever- (allergic rhinitis-) causing tiny plant products that transport the sperm to the egg. Each plant produces huge quantities of such pollen grains, a way to increase the odds that at least one of their many millions will waft its way to a receptive target. Thus, the development of seeds and pollen fit hand-in-glove – and taken together, they enabled plant life to occupy the uplands and, ultimately, the entire land surface. Indeed, seed plants today are arguably the most important organisms on Earth, especially on land where they form the basis of life as we know it.

How did seed plants enter the scene, how did they arise? Again, the fossil record holds the answer. The earliest known seed plant is *Elkinsia polymorpha*, a "seed-fern" dating from the Late Devonian (376-360 Ma) of West Virginia. At first glance, the name of this earliest-evolved group of seed plants, "seed-ferns," seems an oxymoronic absurdity since ferns are spore-plants and seed plants obviously are not. Nevertheless, this odd juxtaposition of "seed" with "fern" is in fact correct. Seed-ferns look like ferns – similar leaves, similar branch habit, similar overall appearance – yet seed-ferns have seeds, ferns have spore-producing spore sacs. But this moniker is also a result of the way fossils are preserved in the rock record, simply because plant parts are typically preserved as individual pieces of a larger plant, not in "organic connection" with the other fossilized remains of the plant. Thus, for years and years, isolated fossil leaves (such as *Alethopteris*, *Neuropteris* and *Macroneuropteris;* **Fig. 3-5)** were assumed to be the foliage of ferns, there being no telling difference between them and those of assured fossil ferns. This misinterpretation was finally corrected when fossil seeds (*Trigonocarpus*; **Fig. 3-6)** were found attached to the mid-veins of *Alethopteris* **(Fig. 3-7)**, a foliage-type known to occur in a large tree-like plant. By that time, both the leaves and the seeds were already well known, but there was no obvious reason to link the two. Finding them in organic connection thus solved a long-standing conundrum and led to recognition of a new fossil plant group, the medullosaleans, named after its best-known example, the seed-fern *Medullosa* (**Fig. 3-8**). Indeed, if you compare *Medullosa* with the co-existing Carboniferous tree fern *Psaronius* (**Fig. 3-2B**) you will see that except for the presence of seeds hanging off the leaves of one but not the other, there is not much difference.

Fig. 3-5 Carboniferous fossil seed-fern foliage, (**A**) *Alethopteris*, (**B**) *Neuropteris*, and (**C**) *Macroneuropteris*.

Fig. 3-6 *Trigonocarpus parkinsoni*, nut-like Carboniferous 1 inch- (2.5 cm-) long seed-fern seeds.

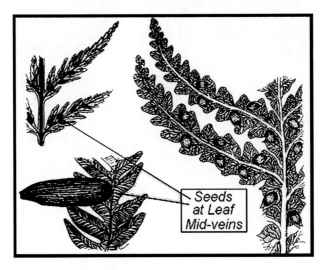

Fig. 3-7 *Trigonocarpus* seeds attached to the mid-veins of Carboniferous seed-fern foliage.

Fig. 3-8 The Carboniferous to Permian seed-fern *Medullosa*, a small to medium-sized tree up to 30 feet (10 m) in height, the largest having *Alethopteris* fronds 21 feet (7 m) in length.

Fossil plants show that South America and Africa were joined

Here the story gets even more interesting because it illustrates how detailed studies of Paleozoic fossil plants can help change perceptions of the evolution of planet Earth and how, again – as in the steps leading to the ultimate acceptance of Zimmermann's Telome Theory, outlined above – well-meaning but misguided purveyors of science by authoritative assertion can influence the advance of knowledge

Step back to an earlier time when Western societies moved through the "Age of Discovery" (dating from the beginning of the 15th century when in the 1400's sea-faring European nations, primarily France, England and The Netherlands, set their sights on exploration and exploitation of the newly discovered American continents, to the mid-17th century when Australia and New Zealand were found to exist). The inquisitive intrepid explorers mapped the coastlines of these previously unknown far-off lands and, quite reasonably, their findings were accepted by the public at large, virtually all assuming that "our land is here, those lands are there, and that's how the world is." Almost no one wondered, *"How did these 'new lands' come to be where they are today?"* Yet a few did and ultimately made a difference. Here's the story of how that transpired.

It is long-established geological lore, handed down from teacher to student over countless generations, that the jigsaw fit between the "elbow" of the eastern coastline of South America and the "arm-pit" of the western coast of Africa was first noted by the English philosopher Francis Bacon (1561-1626) in his 1620 volume *Novum Organum, sive Indicia Vera de Interpretatione Naturae* ("New Organon, or True Directions Concerning the Interpretation of Nature," a tome written in Latin). As historians and Latin scholars have shown, however, though Bacon did note that the shapes of South America and Africa are notably similar, the similarity to which he referred was that of their overall size and shape, particularly the jutting-out of their two eastern-most coasts, both at the "elbow" of South America and at the sharp hump, the "horn" of East Africa. Thus, in actuality, the first to note the jigsaw fit of the two continents was, instead of Bacon, the Flemish cartographer Abraham Ortelius (1527-1598), the mapmaker who in 1596 put together the first modern atlas (*Theatrum Orbis Terrarum,* "Theatre of the World"). In doing so, he noted not only the coastal fit of the eastern margin of South America and the western coast of Africa – the jigsaw fit erroneously attributed to Bacon – but also the approximate fit of the eastern coast of

North America and the western coast of Europe. Sufficiently impressed that this was not pure happenstance, he went on to propose that *"the Americas, Europe and Africa were initially a single land mass."* However interesting this may seem to us, in the1600's it mattered little – Ortelius was "a mere cartographer," not a full-blown learned scholar and, anyway, everyone "knew" that **continents do not move**!

Ortelius' prescient observation therefore languished for a century-and-a-half, finally resurrected – at least for South America and Africa – by the French-Italian paleobotanist Antonio Snider-Pelligrini (1802-1885) in his 1859 opus *Mystères Dévoilés ... L'Origine de L'Amérique* ("Mysteries Revealed ... The Origin of the Americas"). Pelligrini found that the latest Carboniferous- to Permian-age fossil floras of South America, southern Africa and India were all virtually the same, characterized by such Glossopteridalean seed-ferns as *Glossopteris* and *Gangamopteris*. Knowing full well that plants can not fly, swim, or walk, he imagined that these now far-distant locales must have originally been connected. For Pelligrini, that was not enough. Being knowledgeable also about the coeval fossilized flora of the northern hemisphere, fossil plants markedly different from those of South America, Africa and India he had studied in depth, he supposed that although these areas had been part of one giant landmass, this mostly southern megacontinent must have been separate from the lands to the north. He wrote this up, including a nice figure illustrating his view of *avant la séparation* ("before the separation") and *aprés la séparation* ("after the separation"; **Fig. 3-9**). Pelligrini's concept has now been confirmed, the ancient landmass he envisioned known as Gondwanaland (including present-day South America, the Falkland Islands, Africa, Madagascar, India, Australia and Antarctica) and the plant fossils they contained as the Gondwanaland Flora.

Yet Pelligrini's ideas, like those of Ortelius before him were dismissed by the Powers-That-Be. To them, this was an impossible claim made by "a mere paleobotanist." After all, as everybody "knew," **continents do not move**! Indeed, before the ultimate acceptance of Continental Drift-Plate Tectonics, even those well acquainted with the fossil evidence regarded Pelligrini's notion of a giant late Paleozoic landmass as gibberish, pure hogwash – as I discovered in my second-year college paleontology course in 1961, a full century after Pelligrini's seminal finds. To his credit, the prof in the course readily acknowledged the seeming identity among the South American, African and Indian fossil floras. But he, like all other paleontologists at the time explained this away as being a result of the connection of these lands by supposed "land bridges." (At the end of that class session – thousands of mile-long ocean-crossing land bridges being "new news" to me – I ventured up to ask the prof whether there was any supporting evidence. He asked me what I meant. I explained: *"A chain of submerged land bridges or sediments derived from them strewn across the ocean floor."* He said, *"No, the bridges were probably chains of volcanic islands."* When I then asked whether there was any evidence for this, his answer was again *"No."* I left that short conversation thinking that land bridges were some sort of made-up fabrication. The prof. may very well have thought, *"This young upstart asks too many questions."*)

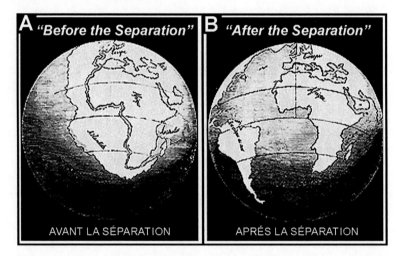

Fig. 3-9 (A) Pelligrini's fossil flora-based illustration of one great southern hemisphere landmass, now known as Gondwanaland – *avant la séparation,* "before the separation," and **(B)** *aprés la séparation,* "after the separation," the landmass having divided to form South America and Africa. An illustration from *Mystères Dévoilés ... L'Origine de L'Amérique* ("Mysteries Revealed ... The Origin of the Americas, Paris, Librairie A Franck and Librairie E. Dentu, 1859).

Though Pelligrini's paleobotanical finds were undisputed, his megacontinent interpretation of them was cast aside. Nevertheless, his basic idea of South America, Africa and India having once been connected into a single giant landmass did not die. The next player in this on-going saga was the German meteorologist Alfred Wegener (1880-1930) who in 1915 formally proposed the concept of "Continental Drift," basing his notion on those who had come before him. Despite being a respected scholar, Wegener was "a mere meteorologist," not a "real" geophysicist, the scientific discipline having the final say about such matters. To the geophysicists, having no interest in or knowledge of the plant fossil record, the basic problem was that Wegener had proposed no mechanism to explain how it was possible, much less even conceivable, that massive continents could actually move across the surface of the globe – a Churchillian riddle wrapped in a mystery inside an enigma.

World War II resolved the riddle

The series of advances that finally set in place our present-day understanding of the underlying mechanisms of such continental movement began in the 1920's when the Germans invented echo-detecting devices. This new technology resulted in discovery of the Mid-Atlantic Ridge, the volcanic mountain range that floors the Atlantic Ocean half-way between the Americas and Europe-Africa and only rarely pokes its head above the surface of the ocean (as it does in southwestern Iceland and at the mid-Pacific island groupings of the Azores and the Ascension, St. Helena, and Tristan da Cunha Islands).

Though the echo-detecting devices were not designed to penetrate deep into the abyss, they did detect the tips of the submerged mountain chain, a huge advance of knowledge. Then, in the 1940's, Donald Griffin (1915-2003) entered the scene, a Harvard biologist studying the behavior of bats and, in particular, how they avoided flying into the walls of the pitch-black caverns in which they lived. He studied their anatomy and, noting their exceptionally large ears postulated that they were sending out sounds that bounced back from the walls, sounds beyond the range of human-hearing. His further studies showed this to be correct. As things turned out – and remarkable as it may seem – bat biology laid the foundation for understanding how Earth's continents have moved over geological time!

World War II then descended. Academia mobilized to help the war effort. My Ph.D. professor, Elso Barghoorn (1915-1984), headed off to Panama to study the fungi befouling the leather binocular pouches of troops in the South Pacific theatre; my father, a paleobotanical coal-geologist regarded as an "essential worker" by the U.S. Geological Survey, was an Air-Raid Warden in Pittsburgh Pennsylvania. Don Griffin contributed, mightily, by perfecting Sonar.

The U.S. was sending supplies, first food and medicine and later, troops, to England, heading out of the northeastern U.S. and then, via Halifax Nova Scotia, across the North Atlantic. But the Germans had an answer: U-boat "wolf packs" of 6 to 8 attacking small submarines. Ships were sunk, thousands of sailors and civilians died. To counter this onslaught, the U.S. employed "depth charges," explosive-filled barrel-shaped casks dropped from the stern of a ship whenever the towed Sonar showed U-boats in the vicinity. Most of the time there were no submarines to detect and the Sonar therefore, for the first time, mapped the topography of ocean bottom and of the massive submerged Mid-Atlantic Ridge. This was new, all thanks to Griffin's bat-based perfection of Sonar.

World War II ended and Europe, Japan and the U.S began to return to normalcy. But by the early 1950's, the US-USSR "Cold War" had commenced. As a response to this development, the U.S. invented deep-diving nuclear submarines able to stay submerged for months at a time. Although ostensibly intended for defensive purposes, this new class of subs was more obviously used for surveillance of the Soviets. To conduct this activity, the subs needed a deep-sea "highway" to get from the U.S. to their listening posts off the coast of the USSR, the biggest obstacle in their path being the 10,000 mile- (16,000 km-) long and nearly 1.5 mile- (2 to 3 km-) high submerged Mid-Atlantic Ridge. The WW II ocean-basin maps showed the path, used for the following decade. The maps were therefore classified as "top secret." All this changed in about 1960 when a Soviet submarine was sighted off the U.S. northeastern coast. With the U.S. military now knowing that the Soviets had also discovered the "secret" undersea highway, the maps were officially declassified in 1961.

At this point, the last major player in this saga entered the scene, Harry Hammond Hess (1906-1969), an American geologist and a United States Navy officer in World War II. Hess was a professor of geology at Princeton University who had become interested in the submarine geology of the oceans while serving in the US Navy, soon becoming Captain of an attack transport ship, the *USS Cape Johnson*, equipped with the then-new technology of Sonar (also called "echo sounding"). At the end of the war, he remained in the Naval Reserve, rising to the rank of Rear Admiral.

Thus, when the ocean-bottom maps were declassified in 1961, Hess had both the geological knowledge and the personal experience necessary to place the maps and the data they contained into a coherent whole. Hess envisaged that ocean basins grew from their centers, with molten volcanic lava oozing up from the Earth's interior to form mid-ocean ridges. This repeatedly created new seafloor, which then spread away from the ridge in both directions, widening the ocean basin as the central volcanic ridge became increasingly higher. As spreading continued, the newly created ocean floor cooled to form the abyssal plain and this, in turn, moved away any landmasses originally adjacent to the widening center. Hess published his theory in 1962 but he well knew, as

Wegener had learned earlier, that his notion was essentially anecdotal, a "nice story" that though it fit the newly available data lacked the telling geophysical evidence needed to convince the skeptical naysayers.

Fortunately for the scientific community, the needed evidence came only a year later. In1963 two British geologists sealed the case, Cambridge University Ph.D. student Frederick J. Vine and his mentor, Drummond Matthews (1931-1997). Their work looked at the patterns of "magnetic stripes" that parallel both sides of the mid-Atlantic ridge across the ocean floor – like Hess' Sonar-based maps a product of World War II when, to augment Sonar, ship-towed magnetometers were used to detect the magnetic signal of the metallic hulls of U-boat "wolf packs." If no U-boats were present to bounce the signal back, the magnetometers recorded the magnetic signal of the ocean seafloor.

Why should such magnetic strips exist? To many, their mere existence seems plenty odd – though it is easy to understand. When molten lava rises up through volcanic mid-ocean ridges and cools it preserves a record of the polarity of the Earth's global magnetic field, principally because minerals in the cooling lava, chiefly the mineral magnetite (Fe_3O_4), are magnetic and thus align with the Earth's north-south magnetic field when they crystallize. The data showed that seafloor rocks were magnetically striped, always in a north-south orientation, and that these stripes systematically varied from wide to narrow in a symmetrical pattern on both sides of the central ridge, the rock-recorded "magnetic needle" pointing north on both sides at similar distances from the central volcanic ridge and then pointing south in the next stripe in the series. If Hess was right, Vine and Matthews reasoned, this symmetrical pattern was no fluke, instead indicating that Earth's magnetic field has switched north-to-south from time-to-time, changing from its current "normal" direction to the opposite "reversed' direction. This idea, like Hess's has been borne out – there being some 183 such reversals recorded in the rock record over the past 83 million years, the significance of the magnetic stripes confirmed by their ages well documented from dating the same magnetic reversals in volcanic rocks accessible on land. In addition, when the volcanic rocks of the seafloor were later dated they were found to be the same age at similar distances away from the ridge on each side. Pretty clearly, Hess's ideas were correct, the ocean floor was created at mid-ocean ridges, the ocean basins widened and, thus, the continental landmasses were shoved to one side or the other.

The crust of the Earth, a thin solid rocky shell that covers the planet's surface, is the top component of the "lithosphere," a division of planet's layers that includes the crust and the upper part of the underlying mantle. The crust itself is composed of two distinct layers, the uppermost being a 20 to 30 mile- (30 to 50 km-) thick continental "felsic" layer (so named because it is relatively rich in the granite-forming minerals feldspar and quartz, the latter composed of silica, SiO_2, and the source of the suffix "–sic" in felsic). The underlying layer is thinner, a 3 to 6 mile- (5 to 10 km-) thick "mafic" layer that is rich in the basalt-forming elements magnesium (Mg) and iron (Fe) that floors the oceans. Due to their differences in composition, the felsic light-colored continental layer is physically less dense, having a specific gravity (S.G.) of 2.7 gm/cm^3 whereas the dark-colored oceanic crust is denser, having a S.G. greater than 3.0 gm/cm^3. Because both the continental and oceanic crustal layers are less dense than the mantle below, both types of crust "float" on the underlying mantle. Formally referred to as "isostasy," this difference in density is also the reason that the continents rise higher than the ocean basins, the less dense continental crust "floating" above the oceanic crust with water pooling across these vast depressions to form the oceans.

Importantly for Hess's theory, isostasy is also the reason that the less dense continental masses move in earthquake-caused fits-and-starts over the underlying more dense oceanic crust. And this, in turn, is the hallmark of what is now generally referred to as Plate Tectonics, Earth's continent-forming uppermost felsic crust being divided into eight giant slowly moving tectonic plates (the African, Antarctic, Australian, Eurasian, Indian, North American, Pacific, and South American) and nine smaller plates **(Fig. 3-10)**.

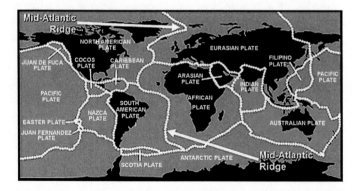

Fig. 3-10 Earth's continent-forming uppermost crust divided into the present-day eight major tectonic plates (the African, Antarctic, Australian, Eurasian, Indian, North American, Pacific, and South American) and nine minor plates (the Arabian, Caribbean, Cocos, Easter, Filipino, Juan de Fuca, Juan Fernandez, Nazca, and Scotia).

Most workers today credit Hess, Vine and Mathews for initiating the concept of Plate Tectonics – and rightly so because they were first put the story together. Pieces of the puzzle, however, have a long preceding history, the most tellingly dating back to Antonio Snider-Pelligrini's Paleozoic fossil flora-based concept of 1859. Perhaps science should have taken more seriously the ideas of this "mere paleobotanist," of his 1596 predecessor the "mere cartographer" Abraham Ortelius and, in 1915, those of the "mere meteorologist" Alfred Wegener. But perhaps not. Most practitioners of science know well that "ideas are cheap," the real cost being the time and effort needed to establish that novel notions accurately describe the real world and another example of the Russian proverb *doveryay, no proveryay*, "trust but verify." Nevertheless, if you pause and think about it, it *is* truly rather remarkable that Pelligrini's Carboniferous-to Permian-age fossil flora – despite the naysayers, despite the incessant drumbeat of purveyors of science by authoritative assertion – could ultimately have had such impact. I'd like to think that now, after such a long period of trial and tribulation, Ortelius, Pelligrini and Wegener would be pleased.

Gymnosperms rule the Mesozoic

By the beginning of the Mesozoic Era (250 Ma), development of the new means of reproduction – seeds and pollen – led to the rise to domination of the land flora by seed plants as they remain today. Of the great varieties of Mesozoic seed plant groups, three major plant groups stand out.

The first of these, the tropical or subtropical **Cycadales**, "cycads," is best known in the modern flora by the "Sago palm," *Cycas revoluta.* Cycads are short and shrub-like, distinguished by their prominent topknot of large leafy foliage and their distinctive thick stems, pockmarked by close-packed pits that denote where the leaves of earlier stages of the plant's growth have been sloughed off **(Fig. 3-11)**. As shown by these densely packed leaf scars, cycads are slow-growers, a single specimen taking many years to mature and fetching thousands of dollars to the nursery where they were raised. Dating from the Mesozoic "Age of Dinosaurs," their thick-walled seeds are a bright red, evidently to entice co-existing herbivorous dinosaurs to eat them and then spread them widely along with their fertilizing excrement. Not surprisingly, the thick stems are easily preserved in the rock record and the prime source of fossil evidence of the group's Mesozoic prominence. Moreover, and unlike their foliage, cycad trunks are especially fire-resistant as has been demonstrated by their survival through the climate change-induced wildfires that raged through southeastern Australia in 2019-2020.

Figure **3-11 (A)** The modern cycad *Cycas revoluta* and **(B)** a Mesozoic fossil cycad stem showing its prominent leaf bases.

The **Ginkgoales**, the second of the three major groups of Mesozoic seed plants, is today represented by only a single species, the "maiden-hair tree" *Ginkgo biloba*. Known well from its Mesozoic fossil record when such ginkgos were widespread and highly diverse, it is rather surprising that the single living example of the group is so well known. This is largely due to that fact that *Ginkgo* is notably resistant to carbon monoxide, a prime effluent of automobile exhaust, and is therefore commonly used as a street tree, for example in New York City. Like many types of seed plants and most animals, *Ginkgo* has separate male plants and female plants (that is, it is "diecious" and thus unlike "monecious" species in which male and female organs occur on the same individual as in birch, oak, pine and spruce trees). Interestingly, however, virtually all the Ginkgos you will ever see are males, not females. This is because the seeds of *Ginkgo* are surrounded by a thick, fleshy, foul-smelling tissue, the "exocarp," a characteristic – like the bright-colored seeds of cycads – that was evidently an inducement to dinosaurian reptiles to feed on them and spread the undigested seeds far and wide. Described by Missouri Botanical Garden and University of Connecticut paleobotanist Henry N. Andrews (1910-2002) as having the acrid smell of "rotten dog vomit," if some unwary walker tracks the remains of these exocarps into a home or a classroom their stench can raise havoc. Thus, commercial *Ginkgo*-producers practice infanticide, destroying the female plants before they mature.

The third major group of Mesozoic seed plants was the **Coniferophyta**, "conifers," seed-cone-bearing "Christmas Trees" like the modern pine tree (*Pinus*), the monkey-puzzle tree (*Araucaria*) and the California redwood (*Sequoia*). Among all of the groups, the conifers, the prime components of Mesozoic forests represented by a great many varieties were the most successful, even today providing the basis for the logging industry. Among the many sites where their fossilized remains can be viewed and enjoyed, perhaps the best known is Petrified Forest National Park in east-central Arizona where their Triassic fossil stems along with those of other conifers are spread across the landscape.

Of these three once highly successful Mesozoic groups, cycads are today represented by a scant nine living genera; *Ginkgo* and its allies by only one living genus and one species; and conifers are a moribund group, their diversity rapidly dwindling. Once again, the demise of these once-successful plant groups is result of evolutionary advance as their descendants, flowering plants, typified by having better protected seeds than their forerunners, came to dominate the flora.

Cycads, *Ginkgo* and conifers are seed plants, collectively known as *gymno*sperms (from the Greek *gymnos*, "naked," ancient Greek wresters battling in the nude, *gymnos* being carried over to the modern term "gymnasium," plus *sperma*, "seed") whereas flowering plants are *angio*sperms (from the Greek *angeion*, "case" or "casing," and *sperma*, "seed"). In other words, gymnosperm seeds lack a protective casing whereas in flowering plants the developing seeds, the "ovules," are encased and thus better protected. Both gymnosperms and angiosperms are seed plants and both rely on pollination for ovule fertilization. What is the protective casing that is present in angiosperms but absent from gymnosperms?

To understand this important difference, we now turn to the studies of Harvard botanist Irving Widmer Bailey (1884-1967), to me a close friend, my professor's professor who occupied the laboratory next to mine when I was a graduate student. (I warmed Prof Bailey's hands each morning when he came in; I darned his socks, as my mother had taught me as a grade-schooler; he watched each day as I tried to teach our building janitor, Willie Casterlow, how to read; and I was an honorary pallbearer at his funeral). Most importantly, it was I.W. Bailey who in 1943 was first to explain the origin of the embryo-enclosing and –protecting "carpel" of flowering plants, a structure characteristic of angiosperms that Bailey called the "conduplicate carpel" that is absent from the naked-seed gymnosperms.

According to prof. Bailey's idea, the angiosperm embryo-enclosing and -protecting carpel resulted from the enclosure of the seed in a folded leaf. To establish this, on a botanical fieldtrip to Fiji in 1942 he and his colleague Albert Charles Smith (1906-1999), a professor of botany at the University of Massachusetts, discovered the new plant *Degeneria*, a member of the early-evolving Magnoliales (the Magnolia tree group). Their new find showed Bailey's concept to be correct, the embryo-protecting carpel was indeed the result of the in-folding of a surrounding leaf **(Fig. 3-12).**

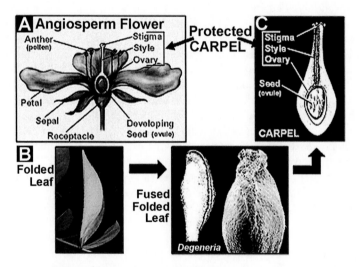

Fig. 3-12 Illustrations showing **(A)** the components of a typical angiosperm flower, and **(B)** the origin of the conduplicate carpel derived from a surrounding in-folded leaf like that of *Degeneria*, a member of the early-evolving Magnolia group and **(C)** the stigma-, style-, ovary-, and seed- (ovule-) protecting carpel.

In retrospect, at least, this was to be expected. After all, the petals that encircle the carpel in a flower are quite obviously pigmented leaves – compare them with foliage of the plant and have a look at their vein structure – as is the ring of sepals immediately below them **(Fig. 3-12).** Why did the origin of the carpel make a difference?

As we have seen before in the on-going evolutionary battle to survive and thrive, the contest to reproduce and propagate one's kind is paramount (and not only in plants, but in animals as well). Yet in all organisms, there is one stage in the life cycle that is the most vulnerable, that part of their development that is most susceptible to the damage that would threaten the continuance of the species. In plants (and in animals also) that stage is the fragile developing embryo. Gymnosperms have naked, unprotected seeds, exposed on the upper surfaces of their cone scales to the whims of the local environment. For angiosperms, such vagaries have been defeated by enclosing the developing embryo in a stout enclosing sheath, Bailey's conduplicate carpel.

How old are the angiosperms, when did they arise? To Darwin, the answer to this question was uncertain as he wrote to his botanist friend Joseph Hooker in 1879, referring to it (or, more properly, to the origin of advanced dicotyledonous angiosperms) as an *"abominable mystery."* And so, this quandary has largely remained, mostly because of the incompleteness of the early fossil record of the group. Nevertheless, from time-to-time new finds of purported Mesozoic angiosperms have been recorded. In 2015, for example, an international team of scientists led by Indiana University paleobotany professor emeritus David Dilcher reported the discovery of the new aquatic fresh-water fossil angiosperm *Montsechia* in strata of the Pyrenees mountain range of northeastern-most Spain. The discovery of these fossils, having affinities to the modern hornwort *Ceratophyllum*, was an important find that extended the established fossil record of flowering plants into the mid-Cretaceous, about 130 Ma ago. Then, in 2018, only three years later, research professor Qiang Fu and his colleagues of the Nanjing Institute of Geology and Paleontology discovered flower-like structures of an even older possible angiosperm, *Nanjinganthus*, in mid-Jurassic 174 Ma-old sediments in the Nanjing region of eastern China. Because age-estimates based on genomic phylogenies suggest that the flowering plant lineage originated about 140 Ma ago, not nearly as early as the mid-Jurassic, and because the exact affinities of *Nanjinganthus* are uncertain – the fossils reported to exhibit the characteristics of angiosperms generally, not of a specific flowering plant or angiosperm sub-group – this 2018 report has engendered a degree of skepticism.

Thus, all one can say at present with certainty is that angiosperms date from the Mesozoic, having arisen at least as early as the mid-Cretaceous and, possibly, appreciably earlier. Darwin's "abominable mystery" remains incompletely solved, but progress continues to be made.

The K-T Chicxulub Impactor

For many of us, the termination of the Mesozoic and the beginning of the Cenozoic, some 65 Ma ago – known also as the Cretaceous-Tertiary (abbreviated "K-T") boundary – is marked by the extinction of the dinosaurs. And. indeed it is, though the fossil record shows that by that time these giant reptiles were "on the way out," their diversity gradually diminishing over the preceding hundred million years. Moreover, the demise of the dinosaurs is only a part of the story, their extinction being only a fraction of a worldwide "mass extinction event" that marks this Era-changing transition in the history of life.

The basic cause of this worldwide catastrophe was the impact of a massive meteor or planetesimal body, attracted by Earth's gravitational field, crashing down offshore the Yucátan peninsula of eastern Mexico in the Gulf of Mexico. As documented in the rock record, this interplanetary body – dubbed the "Chicxulub Impactor" after the nearby village of Chicxulub – was enormous, some 6 miles (10 km) to perhaps as much as 50 miles (81 km) in diameter. The product of this we see today is the impactor's crater, more than 93 miles (150 km) in diameter and one of the largest known impact craters on Earth. The effects of this impact, releasing some *two million* times more energy than the most powerful nuclear bomb ever detonated, were devastating, not only locally but globally and long-term, for months on end.

The total planet "rang like a bell" as impact-triggered seismic shock waves shook the planet and released pent-up energy sequestered in Earth's crust, causing earthquakes and volcanic eruptions. Impact-caused tsunamis, literally miles in height, surged out of the Gulf of Mexico to the interiors of continents drowning many forms of land life. A considerable thickness of sediments rich in the mineral anhydrite ($CaSO_4$) formed part of the Yucátan target area. The high heat resulting from the impact thus drove off gaseous sulfur dioxide (SO_2) from the anhydrite, releasing hundreds of billions of tons of sulfuric acid aerosol into the atmosphere. This aerosol, in turn, caused a cooling of Earth's surface – as it does today when it is emitted from volcanic eruptions – decreasing photosynthesis by orders of magnitude and depleting the ozone layer and its blockage of life-threatening UV radiation.

As the planet cooled the skies then released a long-lasting deluge of sulfurous aerosol-derived acid rain that devastated whatever remained of life on land, lacustrine and near-surface marine ecosystems. Moreover, and unimaginably ruinous to the land flora, the heat generated by the impact broiled Earth's surface and ignited wildfires worldwide. The resulting soot and debris clouded the atmosphere, continuously darkened the skies and soon plunged the entire planet into a prolonged total blackness that lasted for months, possibly years. To those acquainted with the USSR-USA Cold War – coming full-blast in the late 1950's and '60s when grade-schoolers were taught to "duck and cover" – this seems eerily similar to the "Nuclear Winter" scenario envisioned to follow an all-out global nuclear holocaust. In actuality, however, the effects of the K-T Chicxulub impact event were far more widespread and far more devastating than even the worst nuclear catastrophe that any of us could ever imagine.

Because the trees and shrubs on land had been wiped out, the large plant-feeding dinosaurs, herbivores, died within weeks from lack of food, a devastation magnified by the precipitous fall in global temperature and the lack in such cold-blooded (ectothermic) reptiles of internally governed body-temperature control. Because the herbivores died, the meat-eating carnivorous dinosaurs died as well a month or two later. Overall, the loss of biodiversity was incredible, literally unbelievable. Virtually the only survivors on land were small scavenging mammals that could burrow into the ground and emerge to eat whatever remained.

The landscape was soon depopulated, desolate. Even the ocean surface layers were affected, an estimated 90% of near-surface phytoplankton soon exterminated opening the upper reaches of the oceans to the massive proliferation of blooms of ancient resilient cyanobacteria, marine "pond scum," that served as fodder for whatever shrimp and shrimplike planktonic krill had escaped the catastrophe. Indeed, the only habitat to escape the devastation was the deep-water world and a few shallow-water settings where the inhabitants were able to find refuge below this worldwide catastrophe, evidently a prime reason that crocodilians, able to hibernate for years at a time, managed to get through the crisis to become the only large reptile to survive and the dominant predator of today's tropical or subtropical near-shore environments.

All told, the K-T mass extinction killed off a fossil-record estimated 80% of animals, 60% of land plants and 70% of all species on Earth. Indeed, the great majority of land plants to carry though to the Cenozoic were small ground-hugging angiosperms pre-adapted to living through and growing rapidly after periods of low light intensity and climate change. The giant trees, mostly forest conifers, died out.

How is it that we know all this? Clearly, the preserved fossil record sealed the case, but how did the fossil-hunters, the paleontologists and paleobotanists, know to find the telling evidence? Although the basic outlines of the K-T extinction event have been known since William

Smith's proposal of The Principle of Faunal Succession in the early 1800's **(Chapter 1)** and are included in Darwin's 1859 opus, understanding of the cause of the event and its globally devastating effects are of far more recent vintage. In fact, this great leap forward dates from the 1980 studies of Luis Alvarez (1911-1988), a Nobel Prize-winning physicist, and his son, geologist Walter Alvarez (University of California, Berkeley) who discovered a centimeter-thick layer of iridium-enriched clay at the K-T boundary. They knew, as does the geological community, that iridium is rare on Earth but abundant in meteorites and asteroids. The Alvarezes published their findings in 1981, postulating that this thin layer of iridium was deposited following the impact by a large meteor-like object that smashed into the Earth. Per usual, this new idea was pooh-poohed … "I never heard that before, it must be wrong." Yet Luis Alvarez was a Nobel Prize-winner, highly respected in the scientific community and unlikely to be mistaken. Over the ensuing decade, other scientists began finding layers of iridium at various places around the globe that seemed to corroborate the Alvarez theory. There was, however, no smoking gun in the form of an impact site. That, however, came in 1991, only a decade later, when the Chicxulub impact crater was discovered off the coast of the Yucátan peninsula.

Angiosperms dominate the Cenozoic

In the Cenozoic (65 Ma to the present), flowering plants came to the fore, out-competing virtually all other components of the land flora, in no small measure aided by the rise of bees and other flowering plant-pollinators (such as beetles, butterflies, moths, humming birds, even bats). This plant-animal interaction, a decidedly more efficient means to distribute pollen to potentially receptive seeds than the broadcast wind-borne pollen of their gymnosperm forerunners, permitted flowering plants to invest more of their energy in plant growth, less in their means to reproduce. The pollen-spreaders benefited as well, visiting flower after flower to sip the sugary (glucose-rich) nectar pooled deep within a flower, the plant-needed pollination being a byproduct of the spread of pollen grains stuck to the pollinators' bodies during the repeated visitations. By this time, flowering plants had diversified into the two great groups of angiosperms we know today, the "dicots" – most of the trees and shrubs around us – and the "monocots," including a few trees (for example, the coconut palm, *Cocos nucifera*), the grasses (members of the plant family Gramineae) and some grain-producing plants (such as corn, *Zea mays*, and rice, *Oryza sativa*). Why are dicots and monocots the two major divisions of angiosperms?

How do they differ and what is meant by the name ending "-cot" they share?

Dicot is a shortened form of the technical moniker of this subdivision of flowering plants, "dicotyledonous angiosperms," and monocot, that for "monocotyledonous angiosperms." The question thus reduces to "what is a cotyledon?" The answer is simple. The cotyledon is a "seed-leaf," the leaf that first emerges from a germinating flowering plant seed – dicots have two whereas monocots have one. As shown in **Fig. 3-13**, the two groups differ in other traits as well, most obviously the number of petals in their flowers (dicots have petals in 4's and 5's, monocots in 3's and 6's) and the branching pattern of the veins in their leaves (dicots have branching veins, monocots have parallel veins). There is uncertainty as to which group evolved first but most workers agree that both were present in the mid-Mesozoic, 200 million to 160 million years ago, and that monocot grasses first appeared near the K-T boundary, about 65 million years ago.

Fig. 3-13 Cotyledons are seed-leaves, the leaves that first emerge from a germinating flowering plant seed, **(A)** two seed-leaves in dicotyledonous angiosperms (dicots) and **(B)** one leaf in monocotyledonous angiosperms (monocots). Similarly, dicots **(A)** have flower petals in groups of 4's and 5's, monocots **(B)** in 3's and 6's, and dicot leaves **(A)** have branching veins, monocot leaves **(B)** have parallel veins.

In sum, to life's evolution the newly "emptied habitat" presented by the Chicxulub Impactor-caused K-T mass extinction event presented an unprecedented opportunity. Flowering plants, previously relatively subsidiary members of the land plant flora rose to dominance, as did mammals, previously out-competed by the far larger and far more numerous reptiles. This, then, led to the life we know today – a world dominated by flowering plants and mammalian animals. And we know that world well, not only because we are part of it but because of the constant erosion and recycling of Earth's rock record (the "geological cycle") that has resulted in the evidence available from these relatively young rocks and the fossils they contain being vastly more plentiful and better preserved than those of older geological strata.

How plants changed the planet – an overview

As we have seen in the last two chapters, the evolution of plants did indeed change the world. Here then is a synopsis, a ten-point summary of the history of plant-life on land.

(1) The earliest plant-like life forms to occupy the land surface were early-evolving cyanobacteria – primitive photosynthetic microbes – dating from well before two billion years ago. The prime constructors of shallow-water intertidal fossil stromatolites and their modern counterparts, "microbial mat communities," they inhabited a tide-governed daily-changing wet to dry environment.

(2) This adaptation to an alternating wet-dry habitat permitted cyanobacteria, accompanied by similarly stromatolite-derived anaerobic and aerobic bacteria, to occupy the dry land. Later, perhaps as early as 700 million years ago, their well-established biota was joined by stromatolite-derived originally aquatic algae and fungi, an advance that set in place the underpinnings of the modern producer-consumer-decomposer continuously recycling land ecosystem.

(3) As early as the mid-Silurian, about 440 million years ago, vascular plants invaded the land environment. First evidenced in the fossil record by *Cooksonia* **(Fig. 2-6)** and best known from the Devonian-age flora of the Rhynie chert, the earliest of these were slender short rhyniophytes **(Fig. 2-2)** replete with the innovations required to have made the transition – green chlorophyll-bearing naked stems, root-like ground water-absorbing rhizomes, tubular water- and food-conducing xylem and phloem, stems coated with waxy cutin to retain the transported water, open-shut stomates to permit inflow during daylight of photosynthesis-sustaining CO_2, and spores that like the stems were cutinized to protect them from desiccation.

(4) Lacking leaves, the rhyniophytes were close-packed **(Fig. 2-3B)** and although limited to near-shore marshy settings by their moist-surface-requiring spore-based reproduction **(Fig. 3-1)** their invasion of the land was an enormously successful advance that led to a long-term, literally hundreds of millions of year battle for dominance among their evolutionary successors. Among zosterophylls, the earliest of these descendants were plants such as *Sawdonia* characterized by spiny stems **(Fig. 2-8)**. These, in turn, gave rise to lycophytes such as "ole star-wood," *Asteroxylon*, having stems densely packed with helically arranged small leaves, microphylls featuring a leaf defining and supporting xylem-phloem central mid-vein, the earliest true leaves **(Fig. 2-9)**.

(5) By the beginning of the Carboniferous, lycophytes had risen to dominate the mid-American Coal Swamp Flora, best known from the fossil scale-tree *Lepidodendron* **(Fig. 2-11)** and its distinctive sprawling shallow root system, *Stigmaria* **(Fig. 2-12A)**. Lycophytes, "spiral plants" characterized by the helical arrangement of their leaves, rootlets and spore cone-components, soon gave rise to the sphenophytes, "disc-plants" that exhibited ridged hollow stems from which emerged, at regularly spaced ring-like nodes, disc-like whorls of branches or of single-veined microphylls, the stems terminating in spore cones. Best known from the fossil record by the Carboniferous sphenophyte *Calamites* **(Fig. 2-13)**, the sole member of the group surviving to the present is the horsetail plant *Equisetum*.

(6) The single-veined microphylls of lycopsids and sphenopsids was an important step forward from their leafless predecessors, allowing them to grow taller, stouter and faster than their competitors. Nevertheless, what was needed in the never-ending competition to dominate the landscape was a larger better-supported foliar organ that being bigger and flatter could better soak-up sunlight, a large robust multi-veined planar megaphyll that, as shown by the fossil record, evolved from rhyniophytes via the Telome Theory **(Fig. 3-3)** and was earliest exhibited by fossil ferns.

(7) Megaphylls by themselves being insufficient to enable marsh-dwelling spore-plants to transition to the dry land surface, a new means of reproduction was needed, one independent of the need for a moist environment to reproduce. This need was met by the Carboniferous origin of seed-ferns characterized by fern-like foliage that reproduced by seeds and wind-born pollen **(Figs. 3-5 to 3-8)**, not by spores and their water-requiring swimming sperm. Such seed-ferns, now extinct, are the group that links ferns to all subsequently evolved megaphyll-bearing plants. Dominant members of the Permian Gondwanaland flora, seed-ferns played a pivotal role in development and acceptance of the concept of Plate Tectonics **(Fig. 3-10)**.

(8) With the innovations of seeds and accompanying pollination, subsequent lineages – most notably cycads **(Fig. 3-11)**, ginkgoaleans and conifers – rose to dominate the flora, making the Mesozoic Era "The Age of Naked Seed Plants," so named because the seeds of such gymnosperms lack the embryo-encompassing and -protecting additional tissue layer of the carpel of their descendants, the angiosperm flowering plants that dominate the flora today.

(9) The virtually unbelievable catastrophic effects of the Mesozoic Era-ending Chicxulub impact event "cleaned" the landscape, opening new opportunities for evolutionary advance. Angiosperms capitalized, many of their prior gymnospermous competitors having become extinct – as did mammals, their previous prime competitors, the reptilian dinosaurs having similarly perished. The two main groups of angiosperms, dicots and monocots **(Fig. 3-13)**, rose to dominance.

(10) Of all of the ancient fossil plant groups, few have survived to the present. The modern flora includes no rhyniophytes or zosterophylls, only one sphenophyte (*Equisetum*), just two lycophytes (the club-mosses *Lycopodium* and *Selaginella*; a single species of the Ginkgoales (*Ginkgo biloba*), and only nine genera of cycads. So it is in evolution, the less adapted lineages dying out, replaced by their more successful descendants.

All this taken together, the history of Phanerozoic plant life can be divided into three great stages: The Paleozoic "Age of Spore Plants," the Mesozoic "Age of Naked Seed Plants," and the Cenozoic "Age of Flowering Plants" **(Fig. 3-14)**. Spurred throughout all this time by competition to grow faster, stouter, taller and to reproduce increasingly more effectively, the plant world has arrived at its current status. As you look into the future, you can bet your bottom dollar that over the millions of years to come, plants will become even better and better adapted to the world around them.

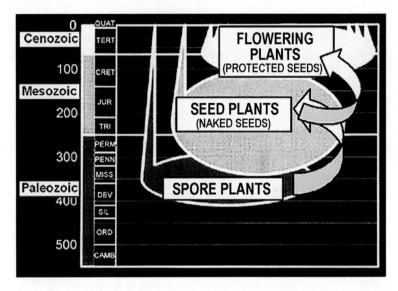

Fig. 3-14 Simplified overview of the evolution of land plants, from Paleozoic spore plants to Mesozoic naked seed plants (gymnosperms) to Cenozoic protected seed flowering plants (angiosperms).

CHAPTER 4

THE CAMBRIAN EXPLOSION OF LIFE

Boris V. Timofeev sets the stage

The rise of animal life is marked in the geological record by the beginning of the Cambrian Geological Period, roughly 550 Ma ago. As is noted in **Chapter 1**, however, this truly remarkable event – mind-boggling in its impact on the evolutionary development of life on Earth – was preceded by a series of major stage-setters, most importantly the advent of oxygen-producing photosynthesis some 3 billion years ago and of eukaryotic, nuclei-containing microscopic cells, about 2 billion years ago.

Primitive, early-evolving eukaryotes – relatively large microscopic single-celled spheroidal algal phytoplankton referred to as "acritarchs" from the Greek *ákritos*, "doubtful, indistinguishable," so named because of their uncertain affinities to modern algae – are known from the fossil record soon after Earth's surface-environment had become habitable by oxygen-requiring forms of life. Their appearance about 2 billion years ago was then followed by a billion year-long period, the so-called "boring billion" when life evolved hardly at all, the seas dominated by such slowly diversifying asexual phytoplankton and the land surface ruled by slowly evolving asexual microscopic cyanobacteria.

About 1,000 million years (Ma) ago the development of sexual reproduction changed all this. As discussed in **Chapter 1**, earlier-evolved life reproduced by asexual cloning, each offspring being a genetic carbon copy of the parent, whereas in sexual reproduction the genes from two parents are combined such that the offspring, except for identical twins, differ from each other. As a result, beginning about 1 billion years ago the genetic and thus the biological diversity of living systems markedly increased, as did the speed of evolution. The kingpin to the discovery of this great step forward in life's long history was the Ukrainian Precambrian microscopic fossil expert Boris Vasil'evich Timofeev (1916-1982) – even though Timofeev had no real understanding of what he had uncovered.

Boris Timofeev **(Fig. 4-1A)** – the head of the Precambrian fossil laboratory at the USSR Institute of Precambrian Geochronology in Leningrad – was a kind, well-meaning, intelligent and interesting man, but he was unschooled in biology as I learned first-hand in 1975 when I was privileged to spend two months as a visiting scientist in his laboratory. Almost every day, about noon, he invited me into his office where he would pull out a pint bottle of lab alcohol, two small cordial glasses, and a set of tincture bottles of various flavorings – peppermint, cinnamon, rose, plum, and cherry. We sat, we chatted. One morning I asked him *"How do you decide the size-ranges of the various fossils you have described?"* His answer: *"I divide them neatly by their sizes, 1 to 25 microns, 25 to 50, 50 to 100, 100 to 250, and 250 to 500. It all fits together."* (He did not ask me what I thought – and, being merely a guest, I did not venture an opinion – but this was a geologist's arbitrary solution to a fundamental biological problem. The rules for such matters are well defined, based largely on comparisons with modern counterparts, matters about which Timofeev had no knowledge. His size categories simply did not fit the real world.)

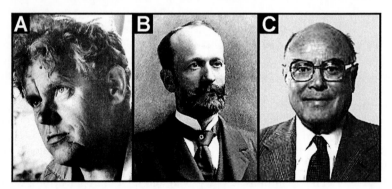

Fig. 4-1 Luminaries discussed in this Chapter. **(A)** Ukrainian geologist Boris Vasil'evich Timofeev, a pioneer in the study of Precambrian microscopic fossils. **(B)** Charles Doolittle Walcott, discoverer of the Burgess Shale Fauna and of the Cambrian Explosion of Life. **(C)** Czech-Australian paleontologist Martin Fritz Glaessner who documented the latest Precambrian soft-bodied fauna of the Ediacaran Period of geological time.

At one of these confabs toward the end of my visit – well after Timofeev had come to know me and trusted me – I asked him to tell me about his past. He paused, pointed up to the ceiling, cupped his ear, and then silently raised both hands and crossed their two first fingers. I had no idea what he meant. He then gestured – again silently – for me to follow him out of his office. The two of us then walked cross the nearby bridge over the Neva River to a public park.

Sitting together on a bench in the park, he then told me about his past – truly trusting me and knowing that I would not mention any of what he had said to the "keepers" (the KGB-dominated Foreign Office) at the Institute. He was a Ukrainian and therefore distrusted by the Soviet authorities (the reason he had pointed to his office ceiling … perhaps the KGB was listening?) and he had been imprisoned behind bars in a Gulag labor camp in Siberia (the explanation for crossing his fingers). After a stint in the Gulag, he was released and assigned to lead a new laboratory at Leningrad because he was the only experienced scientist in the USSR capable of using mineral acids to extract microscopic fossils from shales, crucial to the petroleum industry to determine the age of the fossil- and potentially oil-hosting sediments. He was prohibited from departing the USSR, for fear by the authorities that he would not return (and he confided to me that the authorities were correct, that if he ever *"got out"* he would *"certainly never return"*).

To me, Boris Timofeev's plight was not surprising. This was toward the end of my 6-month visit as a U.S. National Academy of Sciences exchange scientist with the Academy of the Soviet Union – the only American "on the loose" in the entire country – and I was thus trailed daily by the KGB. I ignored them. Each Monday morning at the institute an officer from the Foreign Office (Zinaida) visited me to inquire –though she already knew in detail, from my omnipresent KGB trailers – *"How did you spend your weekend?"* I had nothing to hide. Nevertheless, four different colleagues who had invited me to dinner at their homes to meet their families were interrogated by the KGB, unpleasant encounters that sometimes lasted for hours, simply for having *"that American"* in their homes. Unlike Moscow and Akademgorodok (near Novosibirsk, Siberia) where I had spent the previous four months (and had been vouched for, "protected" by prominent USSR Academicians, A.I. Oparin in Moscow and B.S. Sokolov in Siberia), in Leningrad I was on my own and the KGB was pervasive. This surveillance continued throughout my stay, but I was not the only object of their attention. During the last week of my visit, the 1975 International Botanical Congress was held in Leningrad – but Timofeev and all other workers in the institute were instructed not to attend so that they would not "mingle" with the several thousand foreign scientists (and would thus avoid surveillance of the omnipresent KGB). The scientists complied – they knew the rules – but their absence was a shame, the object of such gatherings being to promote international friendship and cooperation.

Despite Timofeev's lack of biological training, his acid-maceration technique worked excellently, particularly as refined by his star lab worker Tamara Nikolaevna Hermann (in pre-World War II days, "Tamara Nikolaevna German"). Moreover, and unlike the other workers who adopted their technique, in addition to investigating known oil-bearing Phanerozoic sediments they also explored Precambrian shales in an effort to find organic-rich rocks and thus, possibly, untapped petroleum reserves. Dating from the1950's they continuously hit pay-dirt, perhaps best exemplified by their later discovery of the fossil flora of the 1,020 Ma Lakhanda Formation of the Uchur-Maya Region of southeastern Siberia. Indeed, even to this day the Lakhanda shales are well-known for containing the one of the earliest recorded highly diverse phytoplankton-(acritarch-) dominated biotas ever discovered **(Fig. 4-2)**. Neither Timofeev nor Hermann understood the biological affinities of the assemblage – rather than identifying them as phytoplankton they regarded them to be spores of land plants, a lineage then and even now known earliest from rocks some 600 million years younger.

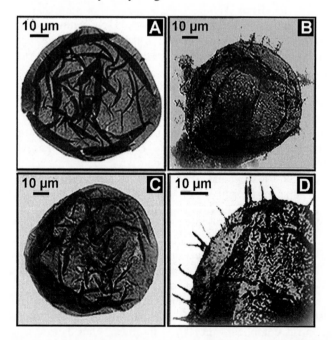

Fig. 4-2 (A, C) Fossil eukaryotic phytoplankton, the spheroidal acritarch *Kildinella* and **(B, D)** the spiny acritarch *Trachyhystrchosphaera* from the 1,020 Ma Lakhanda Formation of southeastern Siberia.

Moreover, as fate would have it and as I had learned in my graduate-school days, this glaring mistake was compounded, magnified by the then-raging East-West "Cold War," the ideological battle between Communism and Capitalism. For example, and though my doctoral professor, Elso Barghoorn, was aware of Timofeev's work, he refused to reference any of Timofeev's publications – not because of his egregious misidentification of the fossils but because, in Barghoorn's parlance, Timofeev was a *"dirty Soviet"* and, thus, a *"Godless Commie."* Others followed suite and the Timofeev-Hermann breakthrough discoveries received scant attention, even in the Soviet Union.

At the end of the day, Timofeev's and Hermann's ignorance of biology, Barghoorn's geopolitically biased predilections, and none of the other distracting aspects of this episode have turned out to matter. The facts have established the case. Spurred by the Timofeev-Herrmann discoveries of shale-compressed fossil microfloras, life's late Precambrian history has become better and better known. Over the half-billion years following the Lakhanda fossils, Earth's biota diversified, evolution rapidly accelerated, eukaryotic multicelled algae and protozoan animal-like protists appeared, all in striking contrast to the immediately preceding "boring billion." For their stage-setting role in understanding this huge evolutionary advance, the immediate prelude to the rise of Phanerozoic life, Boris Timofeev and Tamara Hermann rightly deserve great credit.

Charles D. Walcott discovers the Cambrian Explosion

All scholars interested in life's history are aware of the "Cambrian Explosion of Life." Indeed, the occurrence of trilobites in Cambrian-age deposits was known in the early 1800's when it served as the basis for defining the lowermost entry in William Smith's fossil-based Geological Time Scale **(Chapter 1)**. Thus, the division of Earth's history into pre-trilobite time (the Precambrian) and the time of large life (the Phanerozoic Eon) was established several decades before the 1859 publication of Darwin's *Origin*. Nevertheless, it was not until a half-a-century later and the pioneering studies of Charles Doolittle Walcott (1850-1827; **Fig. 4-1B**) that the ramifications of this major milestone were finally set in place. Here's the story.

As a youth, Charles Walcott attended Utica High School in upstate New York from which he departed at the age of 18 – two years before he was to graduate and after only 10 years of formal schooling (a situation not uncommon in the late 1800's). He then became a farmhand, toiling the fields (and. in that process, augmented the impressive collection of Cambrian trilobites he had begun to amass as a youth). In 1876 Walcott was hired by N.Y. State Geologist James Hall Jr. (1811-1898), by all accounts an irascible tyrant who is reported to have been hoping to acquire Walcott's trilobite collection. (If so, Hall was to be disappointed as Walcott later was to later sell his collection to Harvard's Luis Agassiz Museum of Comparative Zoology). After only two years Walcott departed Hall's employ and in 1879 was recruited to join the newly formed U.S. Geological Survey (USGS), signing on as its third appointee at the munificent salary of $12 per month.

Soon thereafter, the recently appointed USGS Director, John Wesley Powell (1834-1902) set in place a series of expeditions down the Colorado River through the Grand Canyon of Arizona, each a harrowing passage down the rushing waters on large rowboats and unstable log rafts. Walcott eagerly signed-up. On one of the early ventures, in 1883, he discovered the first stromatolites known from the Precambrian rock record, finely layered microbe-produced mound-shaped structures much like the Cambrian "algal reefs" at Saratoga Springs, N.Y. Hall had previously named *Cryptozoön* (meaning "hidden life"), structures that Walcott had examined in 1878 that are now known to occur throughout the Precambrian back to at least 3.5 billon years ago. On a later trip, in 1899, he uncovered another "first," compressed specimens of the Precambrian-age spheroidal planktonic acritarch *Chuaria*. Although Walcott misidentified *Chuaria* as being the flattened remains of a shelly invertebrate animal, "shoe-horning" it to fit with the Phanerozoic fossils he had previously collected, his error made little difference. Given the great age of the fossil-hosting Grand Canyon deposit, the discovery of such an ancient fossil shell would also have been a "first." Today, *Chuaria* stands alone as the first cellularly preserved Precambrian microfossil ever found.

After completing several such trips, Walcott headed north along the spine of the Rocky Mountains and from 1902 to 1907 discovered numerous close-packed stromatolites in the Lewis Range of northwestern Montana ... yet another "first." From there he continued even farther north, up the Rocky Mountains into British Columbia, Canada, where in 1909 he yet again had a huge success, discovery of the remarkably diverse and well-preserved animal fossils of the middle Cambrian (510 Ma) Burgess Shale, the all-important first prime evidence of the Cambrian Explosion of Life.

Like any experienced field geologist, at every locale Walcott made copious field-notes according to one biographer recounting that he spotted the first of the Burgess fossils while he was riding horseback up a treacherous mountain slope and at the base of an escarpment noted a large trilobite-like fossil on a slab of the nearby eroding shale. Perhaps this is true, perhaps not, but his notebook, written each evening around a campfire, makes good reading – he quite obviously was having great fun – and it even includes sketches of the new fossils found that day. The biota he later formally described was exceedingly well preserved, even down to the antennae extending from trilobite snouts, and is wonderfully diverse, including trilobites, sponges, jellyfish, worms and many types of previously unknown primitive animals, an awe-inspiring step forward.

Walcott's career can only be described as truly remarkable, spectacular. After joining the USGS in 1879 he rapidly rose through the ranks, in 1893 becoming Chief Paleontologist and then, in 1894 with the retirement of J.W. Powell, he was appointed the second Director of the U.S. Geological Survey. For Walcott, that was just the beginning. In 1902, he and a team of colleagues met with the exceedingly wealthy philanthropist Andrew Carnegie and established the Carnegie Institution of Washington to promote scientific discovery, a highly distinguished institute even to this day. Walcott later went on to become Secretary (CEO) of Washington's Smithsonian Institution, President of the National Academy of Sciences, and President of the American Association for the Advancement of Science. Additionally, as a close advisor to U.S. President Theodore Roosevelt, he was instrumental in establishing the U.S. National Park system. Throughout all this, beginning in 1909 and continuing for some 17 years until he was well into his 70s, Walcott continued to carry paleontological field investigations of the Burgess fossils. Over the years his finds have been repeatedly confirmed, most recently in 2019 by spectacular discoveries of a great many individual specimens of similarly excellently preserved fossils in the 518 Ma Cambrian-age Qingjiang Biota of central China.

Despite Walcott's myriad accomplishments, not all of his finds were warmly received, most notably his claims of Precambrian stromatolites and cellular fossils. Walcott had ascended to scientific stardom, so it would take another "Biggie" to take him down. Albert Charles Seward (1863-1941), Vice-Chancellor (CEO) of Cambridge University, U.K. and the world's leading paleobotanist, filled the bill. Delivering his verdict from his position of unquestioned authority – Walcott's Precambrian finds being evidence of ancient algae and, thus, under the purview of paleobotany – Seward lowered the boom in his 1931 textbook *Plant Life Through the Ages*, his discussion being brief but to the point. According to Seward, "[Walcott's interpretation of] *Cryptozoön* [the Precambrian stromatolites] *is, I venture to think, not justified by the facts ... It is clearly impossible to maintain that such bodies are attributable to algal activity... we can hardly expect to find in Pre-Cambrian rocks any actual proof of the existence of bacteria."* And Seward's derisive assessment of Walcott's discoveries then concluded with a bit of artful doggerel about Walcott's Precambrian finds: *"Creatures borrowed and again conveyed, from book to book – the shadows of a shade."* Walcott could not offer a rejoinder – he had died four years before Seward's diatribe was published.

Walcott's problem was that the Precambrian fossils he reported were not only previously unknown but were far too old to fit accepted dogma. In retrospect, we can now see that C.D. Walcott had brought the study of Precambrian life to the brink of success, only to have his ship scuttled by Seward's authoritative assertive nay saying. Indeed, as briefly explained in the Preface to this volume, it was not to be until the mid-1960's that the problem posed by Darwin's *"inexplicable"* missing early fossil record was finally resolved. Nevertheless, Charles Doolittle Walcott's pioneering contributions remain highly esteemed. In 1995 the Burgess fossils were featured on the December 4 cover of *Time Magazine* **(Fig. 4-3)**; his pioneering work has been honored by the establishment of the National Academy of Sciences' Charles Doolittle Walcott Precambrian Research Medal; and in 2009, one hundred years after his discovery of the Burgess fossils, Alberta's Big Rock Brewery turned out the Burgess Shale Centennial Celebration Ale. Wow – a *Time Magazine* cover, the foremost international medal in his science named in his honor, and even a celebratory brew!

Fig. 4-3 (A) Four types of Burgess Shale fossils (the carnivorous worm *Ottoia*, the velvet worm *Hallucigenia*, the trilobite *Oryctocephalus*, and the slug-like animal *Wiwaxia*) **(B)** featured on the cover of the December 4, 1995 issue of *Time Magazine*.

Martin F. Glaessner establishes the pre-trilobite, Precambrian Ediacaran

The next advance in this unfolding stage-setting saga dates to 1946 and the work of Reginald Claude Sprigg (1919-1994), an Australian geologist who discovered imprints of fossil jellyfish in strata of the Pound Quartzite in the Ediacara Hills north of Adelaide, South Australia. Sprigg named the fossils *Mawsonites* in honor of the Australian geologist and Antarctic explorer Douglas Mawson (1882-1958), but being uncertain of the fossils' age, Sprigg assumed they were most likely Cambrian.

Beginning in 1950, with the arrival at the University of Adelaide of Czech-Australian paleontologist Martin Fritz Glaessner (1906-1989; **Fig. 4-1C**), the situation markedly changed. By the end the 1950's, Glaessner and his highly talented assistant Mary Julia Wade (1928-2005; **Fig. 4-4**) – who, in fact, did a great deal of the work, discovering, digging out and cleaning the fossils for subsequent publication and public display – this productive duo uncovered firm evidence showing that not only were the Ediacaran fossils pre-trilobite and pre-Cambrian (if barely so) but that the fauna was richly diverse, including many soft-bodied animals previously unknown to science **(Fig. 4-5)**. If Mary Wade was unique, as she certainly was, so also was Martin Glaessner – an internationally known *micro*paleontologist making his biggest findings about *large* multicelled life – and he was always a well dressed "gentleman," wearing a tie and waistcoat even while doing field work.

Fig. 4-4 Mary Julia Wade, the prime co-worker with Glaessner on the Ediacaran Fauna of South Australia.

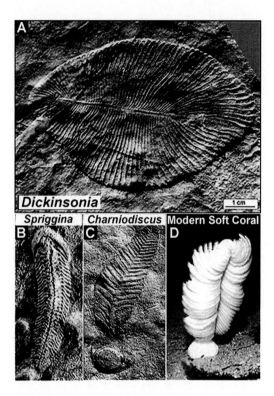

Fig. 4-5 Representative soft-bodied animals of the Ediacaran Fauna of the Pound Quartzite. (**A**) *Dickinsonia*, a disc-shaped worm-like animal. (**B**) *Spriggina,* a polychaete worm-like animal and the oldest known fossil to exhibit an identifiable head, thickened to protect its contained sensory apparatus. (**C**) *Charniodiscus,* a stationary animal that lived anchored to the sandy seabed held in place by a basal holdfast, an Ediacaran fossil similar to (**D**) a modern soft coral.

Like the reticence toward Walcott's earlier break-though finding of Precambrian stromatolites and algal phytoplankton, Glaessner's discoveries were a poke in the eye to traditionalists who were convinced that it was simply not possible for pre-Cambrian animals to exist … *"after all,"* they exclaimed, *"that is the way the beginning of the Cambrian Period of geological history is defined!"*

The leader of this traditionalist school was Preston Ercelle Cloud, Jr. (1912-1991), an influential member of the U.S. National Academy of Sciences (where he is still remembered, in part because he had donated to the Geology Section of the Academy the proceeds of a book of readings he had assembled to fund a cocktail hour at their annual meeting. As things turned out, the book did not sell and the Section received no such largesse. Nevertheless, the "Preston Cloud cocktail hour" has carried on, funded by the contributions of other NAS members).

Pres Cloud had quite a career. After graduating from high school in 1930, he joined the U.S. Navy's Pacific Fleet Scouting Force where he became the fleet's bantamweight boxing champion demonstrating a dominating feistiness – some might say irascibility – which he carried throughout his life. He went on to become the Head of the Paleontology and Stratigraphy Branch of the USGS where, to his credit, he doubled the number of professional paleontologists. At the same time, rather strikingly, he placed the chair at his desk on four-inch risers. When I later asked him about this he explained: *That's true – of course I put my chair on risers – I was the Boss and I had learned years before that underlings would not heed my orders if I did not tower over them.*" Others have suggested that Pres Cloud, rather short in stature, had a "Napoleonic Complex." After leaving the USGS he joined the faculty at the University of Minnesota and, then, the Geology Department at UCLA – each time departing after a short two- to three-year stint and under a cloud (no pun intended) – ending up at the University of California, Santa Barbara.

Ultimately, the Glaessnerian and Cloudian views of the definition of the base of the Phanerozoic Eon came to loggerheads. Glaessner argued that with his new findings of pre-trilobite soft-bodied animals, a geologically older Phanerozoic geological period should be established. Cloud argued the reverse, that no new geological period was needed, that the base of the Cambrian should simply be extended downward whenever new animal fossils were discovered. Clearly, Cloud's idea would not work – each new sequentially older finding would require that the boundary between the Cambrian and the underlying Precambrian be moved to a lower stratigraphic level, a practice that would make it impossible to be sure when the Phanerozoic truly began. At the same time, Glaessner's idea of adding a new Phanerozoic geological period flew in the face of the system used worldwide that was long-established. At the end of the day the controversy was resolved by the International Commission on Stratigraphy and ratified in 2004 by the International Union of Geological Sciences (IUGS), establishing the first new geological period declared in 120 years – but assigned to the Precambrian, not the Phanerozoic.

Both Glaessner and Cloud "won" – Glaessner's new geological period was established, the 635-541 Ma Ediacaran Period – but to Cloud's satisfaction, this new period was relegated to the uppermost Precambrian, not the Phanerozoic. In retrospect, the only scientists who "lost" this battle, an arbitrary definitional matter rather than a dispute over the relevant facts, were Soviet geologist-paleontologist-stratigraphers who years earlier had recognized the Ediacaran segment of geological time as the "Vendian Period." Even now, the Russian "old-timers" remain rankled, regarding this IUGS decision to reflect "unfair international politics" – as it well might have been, given that the debate occurred during the decades-long Eastern Bloc -Western Bloc geopolitical "Cold War."

Fossil record of the earliest animals

Thus far, the story of rise of animal life on Earth has been fairly straightforward: Nucleated cells to sexual eukaryotes, to soft-bodied multicelled animals, to trilobites and the Cambrian Explosion of Life. Is that sufficient? What do we know about the immediate predecessors of multicelled soft-bodied animals? How did they arise? Here the story is not only murky – it is virtually nonexistent. How could that be? Here's the answer.

The precursors of the large many-celled animals of the Ediacaran were no doubt soft-bodied and much smaller, so small that they left no decipherable fossil record. Plausible candidates for such pre-Ediacaran animals include nematode worms, the best-known living example being the "roundworm" *Caenorhabditis elegans* (pronounced Seeno-rabditis ellie-gaans). Because its genus name, *Caenorhabditis* (from Greek *caeno-*, "recent," plus *rhabditis*, "rod-like") is nearly unpronounceable, it most commonly is referred to simply as *C. elegans*. Used by biologists since the early 1960's as a "model organism," its genetics are thoroughly defined showing that it shares many genes and gene-defined molecular biochemical pathways with humans and, therefore, has proved to be a useful model for studying human diseases. The bodies of such nematodes are not divided into orderly regular segments – as, for example, are those of the common earthworm *Lumbricus* – so *C. elegans* is far more primitive than a "real worm."

Roundworms such as *C. elegans* are small, about 1 millimeter in length, and live in the soil where they survive by feeding on microbes such as bacteria. But because they are so tiny they cannot nudge aside stony particles and leave identifiable burrows or trails in their wake as do larger true worms. Moreover, their body wall is thin and soft, too fragile to be readily preservable. Thus, like all other multicelled precursors of the Ediacarian fauna, they have no known fossil record.

What about **their** forerunners, the even more primitive early-evolved animal-like (eukaryotic and heterotrophic) single-celled organisms that led to the emergence of such tiny worm-like creatures? Do they have a decipherable fossil record? Thankfully the answer is "yes," evidenced by a small assortment of fossilized protozoans known from pre-Ediacaran strata. Here the picture is a bit brighter than that of the unpreservable small worm-like animals, primarily because some such protozoans produce encapsulating cysts, thick-walled "resting cells" that enable them to be preserved. Their known fossil record extends some 200 Ma into the Precambrian before the beginning of the Ediacaran, the earliest reported example being the testate amoeba *Melanocyrillium* preserved in carbonaceous cherts of the 800 to 742 Ma Chuar Group of the Grand Canyon, Arizona not far distant from Walcott's 1883 first find of Precambrian stromatolites.

Here, then, is a brief vignette about how *Melanocyrillium* was discovered, a finding that my students and I were lucky enough to make. I had long been interested in visiting the site of Walcott's 1883 seminal find of Precambrian stromatolites (now dubbed "stromatolite ridge"). I therefore obtained the required collecting permits from the U.S. National Park Service, rented a helicopter, assembled a group of three or four of my students and flew in, landing in the vicinity of Walcott's locale where we collected promising rock samples. But the discovery of *Melanocyrillium*, a minute microscopic fossil, was made using optical microscopy after our return – during a collecting trip, when Precambrian rock samples are sampled, one never knows whether they will, or will not bear fruit. After discovering these previously unknown "oddities" and reporting their discovery in a short paper, I turned them over to one of my students, Bonnie Bloeser, who used them to produce a first-rate Master's Thesis (which led to her subsequent highly successful teaching career at San Diego State University).

Protists such as the testate amoeba *Melanocyrillium* played a pivotal role in the subsequent evolution of many-celled animals. The term "protist" includes a grab-bag assemblage of small unicellular organisms – all single-celled and all eukaryotic, some of which are strict autotrophs (like plants), some being strict heterotrophs (like animals), and some of which are capable of switching from one mode of metabolism to the other. Well-known examples of such "switch-hitters" that are commonly shown in high school biology classes include the ciliated heterotroph *Paramecium* and the flagellated photosynthesizing protist *Euglena*.

Cilia, like those in *Paramecium*, are short tiny hair-like fibers that the coat the outer surfaces of a protist and enable it to move across surfaces. In contrast, flagella, such as that in *Euglena*, are much longer thicker whip-like appendages that enable their bearer to swim and, if the protist is heterotrophic, to also lasso food particles from the immediate environment. As documented in the fossil record, it was a particular type of such flagellated protozoans, the choanoflagellates that played a seminal role in the development of Ediacarian animal life.

How could simple single-celled flagellates make such a difference? The seemingly most likely hypothesis is that early in animal evolution the genes of choanoflagellates were transferred into sponges – members of the Porifera (from Latin *porus*, "pore" plus *-fera*, "animal" – "pore animal") – the basal members of the Animal Kingdom. In essence, sponges are large many-celled pore-penetrated water-filled sacs in which some of the pores ("incurrent pores') flush in oxygen-containing water and its entrained food particles and others ("excurrent pores") flush the water and wastes out. According to this theory, the genes transferred from the protozoans to the sponges were fundamental to this flushing, their presence resulting in the production of great numbers of choanoflagellate-like cells known as "choanocytes" that line the interior of the sacs. By whipping their flagella in tandem – like a long chorus line of simultaneously kicking legs – the choanocytes course the water through the sac and extract the nutrients needed to keep the sponge alive.

Sponges are exceedingly simple animals. They do not have nervous, digestive or circulatory systems and lack hard-parts except for small wall-supporting "spicules," tiny multi-pronged bodies that resemble the six-pronged pick-up pieces in the children's game "Jacks and Ball." Indeed, sponges are so simple that the adults are highly regenerative, able to re-form their entire bodies after they have been partly devoured by a predator or damaged in a storm. Moreover, as shown in 1907 by American biologist Henry Van Peters Wilson (1863-1939), they can do this unusual trick of regenerating their body-parts in simple lab experiments. Wilson took adult sponges, squeezed them through fine-mesh sieves onto agar-filled plates, and then watched as the dissociated cells aggregated to reform whole individuals. Try this on a housefly, an earthworm, or even the roundworm *C. elegans* – it will not work!

Knowing the basal position of sponges in the animal world raises one final question, namely, when do sponges appear in the fossil record? Again, the answer is straightforward. The earliest known sponge fossil, *Eocyathispongia*, was reported in 2015 from the Ediacaran Doushantuo Formation, a highly fossiliferous actively mined phosphate-rich deposit in Weng'an County, Guizhou Province, southwestern China. Although much of the diverse Doushantuo biota is permineralized ("petrified"} in the mineral fluorapatite [$Ca_5 (PO_4)_3F$], this earliest known fossil sponge is embedded in limestone (calcium carbonate [$CaCO_3$]) strata of the deposit.

Thus, to fully analyze the fossil, the investigating scientists used mineral acid (hydrochloric acid, HCl) to dissolve the enclosing carbonate matrix and liberate the organic-walled fossil to show its sponge morphology – a good example of the application of Timofeev's acid maceration technique.

Despite the overwhelming evidence of *Eocyathispongia's* Ediacaran age (about 60 Ma before the beginning of the Cambrian) and its definitive poriferan characteristics, yet again – as we have seen before for the seminal finds of Timofeev, Walcott and Glaessner – these "earliest known sponges" hit a snag. As was pointed out to me by David J. Bottjer, one of the authors of the seminal published report, naysayers voiced their concerns: *"There are too few specimens," "soft-bodied sponges cannot be preserved,"* and, the primary complaint, *"they are much too old ... everyone knows that sponges did not exist in the Precambrian."* Despite such negativism, the find held up. Sponges preceded multicelled animals, as is now firmly established by the fossil record. Moreover, the fossil record of sponges continues up the geological column, subdivided into two great groups – those in which their sac-supporting mineralic spicules are composed of calcium carbonate, the "calcisponges,' and those in which they are silica (SiO_2), the "glass sponges."

The animal "Tree of Life"

Given the foregoing we can now see how and when animal life began – from the advent of sexually reproducing eukaryotes about 1,000 Ma ago, to the later Precambrian development of choanoflagellates (the single-celled protist "sister group" of the Animal Kingdom), to the sponges and the other soft-bodied animals of the Ediacaran, to the burst of hard-part containing many-celled animals that marks the Cambrian Explosion of Life. And we have now been introduced to Boris V. Timofeev, Charles D. Walcott, Martin F. Glaessner, Preston E. Cloud and the other "movers and shakers" who set this sequence in place.

Our task now is to ready ourselves for the follow-on discussion of the subsequent evolution of the Animal Kingdom, a fundamentally simple story that includes a number of unexpected twists and turns and a plethora of new concepts and the new terms by which they are known. The entire sequence begins with the development from multicellular (metazoan) sponges of "true metazoans," *eu*metazoan jellyfish, and continues from radially symmetrical jellyfish and corals to bilaterally symmetrical "bilaterians," acoelomate flatworms, and from them to coelomate eumetazoans, animals in which the gut is enclosed in a body cavity **(Chapter 5)**. The eumetazoan coelomate lineage then divides, the story first continuing along the protostome pathway to insects, from coelomate brachiopods and bryozoans to mollusks and then, via the "jointed-legged worm" *Peripatus*, to crabs, trilobites and insects **(Chapter 6)**. We will then trace the other branch of the coelomate tree, the deuterostome lineage, from its base upward leading from sea stars ("starfish") to fish, amphibians, reptiles, birds and mammals **(Chapter 7)**, taken as a whole the evolutionary precursors of eumetazoan deuterostome bipedal primates such as ourselves.

　　　The new concepts – and the new terms – come thick and fast. But to defeat this problem we will use a simplifying roadmap, a stripped-down phylogenic Tree of Animal Life that includes the highlights of the evolutionary progression. Have a look at **Fig. 4-6** and you will see where we have been and where we are going.

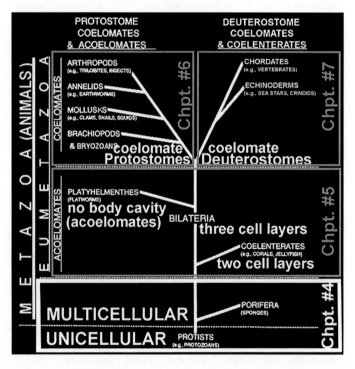

Fig. 4-6 A simplified phylogenetic "Tree of Life" of the Animal Kingdom showing in the white rectangle at the bottom the parts of the tree discussed in **Chapter 4** and, above, those to be discussed sequentially in **Chapters 5 through 7**.

CHAPTER 5

THE RISE OF ANIMAL LIFE

The Animal Kingdom comprises 70% of Earth's biota

The some two million different formally described and catalogued living species are distributed quite unevenly among the various major groups. About 10% are assigned to the prokaryotes (archaea and bacteria, including cyanobacteria) and "lower eukaryotes" (protists and fungi); some 20% are members of the Plant Kingdom; and whopping 70% belong to the Animal Kingdom of which the most diverse by far are insects (an estimated 57% of all living species). These, however, are only estimates, a current best guess of Earth's biota. The "real number" of living species is thought by experts to be closer to nearly five times higher, a total of some 10 million different types of organisms, the vast majority yet to be discovered and formally described.

For the most part, the marked differences among the number of described taxa in the various major groups are easily understandable. We humans, the describers of these species, are relatively large mammals whereas prokaryotes, protists and fungi (except for mushrooms and toadstools), taken together accounting for only 10% of the total, are tiny and not easy for us to detect, much less to describe and name. A great number of plants, as a group representing 20% of Earth's biota, are of course large, some huge, but they don't walk, they don't squawk, so we tend to take them for granted unless they are particularly useful to us. And because we humans are animals, the other members of our animal "tribe," 70% of all named species, are of special interest to us and are thus described and officially named far more frequently.

This leaves only the huge number of insects to be accounted for, about 80% of all animal taxa and a whopping 57% of all named species of life. Why are they so notably diverse? Indeed, there is no question that insects are terrifically varied, with some entomologists estimating that well more than a thousand new species of insects are described and named each year (largely from the South American Amazon rainforest). The reason for this is that insects and their close kin differ importantly from other members of the Animal Kingdom, primarily in the way they are built. Rather than being soft-bodied like an earthworm, or having a strong interior bone structure like us and our mammalian cousins, insects and their close relatives (crabs, lobsters, trilobites and the like) have a stout "exoskeleton," a robust body-covering carapace to which their muscles are fixed on the inside. For small animals, insects in particular, this together with their small size and large muscle mass relative to their body weight is a huge advantage, enabling them to lift and carry objects many times heavier than they are. As we will see in **Chapter 6**, the insect-like exoskeletal body plan has proved to be an enormously successful evolutionary innovation.

Eumetazoan jellyfish and corals

Let's now turn to the subject immediately before us, the post-sponge initial evolution of the Animal Kingdom. The numerous arcane terms used in science can be a bit confusing, so we will begin with a brief reminder of a couple of the definitions introduced in the last chapter. The term **"Metazoa"** refers to *all multicelled animals* and includes sponges – even though they lack tissue layers, a nerve system and virtually all of the other traits of more advanced animals – simply because sponges are composed of many cells (largely, but not entirely, coenocytes). In contrast, *all more advanced multicellular animals* – the subjects of this and of the following two chapters – are grouped together as the **"Eumetazoa"** (*eu*, Greek, meaning "true, genuine," with eumetazoans therefore meaning "true metazoans," a moniker merited because they exhibit the well-defined tissue layers characteristic of the group). Thus, all metazoans, and all eumetazoans are decidedly more complex, more advanced, than the early-evolving sponges.

As we have seen In **Chapter 4**, metazoans (such as *Eocyathispongia*, the earliest known fossil sponge) and eumetazoans (Sprigg's jellyfish and Glaessner's worms and soft-bodied corals) had both already arisen by the Ediacarian, animal evolution then bursting forth in Walcott's Cambrian Explosion of Life **(Fig. 5-1)**. What happened next and how can we order the twists and turns in the follow-on evolutionary sequence? To accomplish this, we turn to the phylogenetic tree of animal life **(Fig. 5-2)**, a simplified summary of animal evolution based on their rRNA phylogeny, biochemical pathways, larvae and embryos, adult tissues and body form, and the known fossil record.

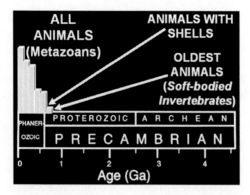

Fig. 5-1 First known appearances of fossilized soft-bodied invertebrates and animals having shells in the geological record.

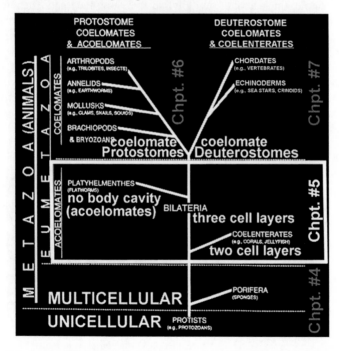

Fig. 5-2 A simplified phylogenetic "Tree of Life" of the Animal Kingdom showing in the white rectangle the part of the tree discussed in **Chapter 5**.

The eumetazoan, post-sponge sequence begins with jellyfish and corals, known collectively as the **Coelenterata** (from the Greek *koilos*, "hollow" plus *enteron*, "intestine," meaning "hollow stomach"), a group also known as "cnidarians." Unlike sponges, which lack tissue layers, and eumetazoans more advanced than coelenterates that have three such layers, jellyfish and corals have two tissue layers, an outer layer, the ectoderm (Greek *ecto*, "outer" plus *derm*, "skin," meaning "outer skin") and the endoderm (Greek *endo*, "within" plus *derm* – "*in*ner skin"). Also, unlike more advanced eumetazoans, coelenterate jellyfish and corals are radially symmetrical cup-shaped organisms that are encircled by a ring of tentacles that spread out from the cup's outer edge.

Jellyfish are gelatinous and transparent – you can almost see right through them **(Fig. 5-3)** – and they are made up of more than 95% water. Given this, and even though some jellyfish are huge, their gelatinous cups (their "medusae") ranging up to nearly 11 feet (3.5 m) across, you might well imagine that there is not much to them. If so, you would be mistaken. For example, think about how jellyfish live, how they eat. Their fodder consists mostly of small floating phytoplankton and zooplankton, the planktonic young stages (larvae) of diverse marine animals, and small fish, worms and shrimp. This they accomplish by use of their long, flexible, easy to see tentacles that are lined with complicated stinging cells, "cnidocytes," each a small compartment that houses a needle-like venom-packed stinger. When some outside force triggers the pressure-sensitive stinging cells – as, for example, when a wandering fish knocks against one of the tentacles – the compartments instantaneously open and fire their venomous dart-like stingers.

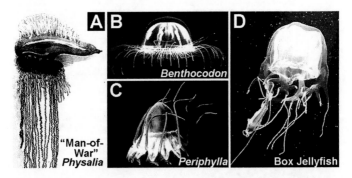

Fig. 5-3 Modern jellyfish (coelenterates), **(A)** the large venomous Portuguese "Man-of-War," *Physalia*; **(B, C)** the more typical and placid *Benthocodon* and *Periphylla*; and **(D)** the deadly "box jellyfish" (known also as the "sea wasp") shown with a captured fish in its grasp.

Though the venom of most jellyfish is not harmful to humans, some can be deadly, for example that of the Indo-Pacific "box jellyfish," so named because of its cube-shaped medusae **(Fig. 5-3D)**. Such jellyfish, known also as "sea-wasps," release a type of venom that makes the heart of a victim contract into a knot. Though modern medicine has devised an antidote for the poison, it acts so quickly that an errant swimmer who has been stung is unlikely to survive without immediate medical attention. Like many marine animals, box jellies are not choosey about their diets, even feeding on other jellyfish (albeit on species different from their own). The encounters with humans are a result of the swimmer's error – the box jellies are simply floating around in hope of ensnaring a small meal (a stung human being far too large for them to ingest).

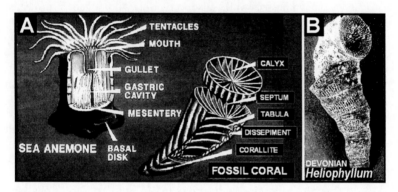

Fig. 5-4 Modern and fossil corals (coelenterates), **(A)** the living sea anemone (members of the coelenterate taxonomic order Actiniaria) and **(B)** the Devonian fossil horn coral *Heliophyllum*.

Why are jellyfish and corals, which look so different from each other, classed together as coelenterates? The answer is simple. Turn a jellyfish upside down and encase it in a rocky calcium carbonate ($CaCO_3$) shell-like husk – voilà, you have coral **(Fig. 5-4)**. Now, pack together side-by-side, cheek to jowl a great many such stony corals (each individual known as a "polyp") and you have a coral reef. Most reef-forming corals have very small polyps but entire colonies, consisting of millions of polyps, grow very large and form massive reefs that can weigh more than a ton. In essence, corals are simply upside-down stay-at-home jellyfish! In other words, unlike jellyfish, corals are sessile, they do not move about. But if they don't move, how can they feed? Again, the answer is simple. Like many other non-motile marine animals, corals are basically "suspension-feeders" – they sit in one spot throughout their lives and use their upward pointing tentacles to snare from the overlying waters whatever food particles fall into their grasp. In comparison with mobile animals, worms and trilobites for example, this feeding technique does not provide a great deal of food. But, then, because corals do not move about they do not need a great deal of food and the energy it provides.

Moreover, many corals have another trick up their sleeve, namely, their tissues house single-celled green algae, "zooxanthellae," tiny cell-enclosed endosymbionts that by their photosynthesis produce nourishment to their animal host. The algae are called *endo*symbionts because they live *within* the coral's tissues and their relationship with the hosting coral is called a mutualistic symbiosis, a "you scratch my back, I'll scratch yours" interaction. The coral's tissues produce the CO_2 and H_2O that the algae need to photosynthesize while also providing the algae a protected environment where the omnipresent feeders outside cannot get in to gobble them up. In return, the algae produce part of the oxygen needed by the coral to breathe and not only help the coral to remove wastes but also supply the coral with energy-rich glucose, the principal product of their photosynthesis. Thus, the coral and the endosymbiotic algae live in harmony, each aiding the other, with as much as 90% of the organic material photosynthetically produced by the zooxanthellae being transferred to the host coral tissue. Colonial, close-packed corals are a common result of just this relationship, a good example being the Devonian coral *Hexagonaria* (the official state fossil of Michigan where they are commonly known as "Petoskey stones").

This all seems fine and dandy, and in normal times it has long been the way that coral reefs have thrived. But the present is not what one would call a "normal" time, primarily because of Global Warming. As atmospheric CO_2 has steadily increased across the planet since the beginnings of the Industrial Revolution in the early 1800s – due to the burning of coal and natural gas, the fossil fuels that provide the energy on which modern civilization has grown reliant – there has been an accelerating increase in worldwide average temperature and that, in turn, has added heat-absorbing moisture to the atmosphere. A prime example of the "Law of Unintended Consequences," the link between this human-caused increase in CO_2, atmospheric water content and climate change was first brought to the world' attention in the 1960s by the climate science pioneer Charles David Keeling (1928 -2005) who from his meticulous monitoring of Earth's atmosphere at the Mauna Loa Observatory on the big island of Hawaii, sent out the alarm.

But Keeling was not the discoverer of this effect. In fact, that dates from a century earlier and the laboratory studies of Eunice Newton

Foote (1819-1888; **Fig. 5-5**), a ground-breaking American botanist and chemist from Bloomfield, New York who in 1856 established that atmospheric CO_2 and water vapor trap heat. Her seminal findings were reported at that year's annual meeting of the American Association for the Advancement of Science, but not by her – although in principle female scientists were reportedly permitted to address the AAAS. Instead, she turned her findings over to a male colleague, the Secretary of the Smithsonian Institution Joseph Henry (1797-1878), a highly respected physicist and fellow Upstate New Yorker who presented them in her stead. Therein lies a useful lesson. Think about it for a moment: Important findings that the discoverer was disinclined to publicly present at a national scientific meeting evidently because of gender inequality? That would not happen today! And that, in turn, illustrates the progress that has been made, both in the scientific community and the society at large, particularly over recent decades.

Fig. 5-5 Eunice Newton Foote, botanist and chemist who discovered that atmospheric CO_2 and H_2O vapor trap heat, the basis of Global Warming.

Nevertheless, and despite our well-meant (if half-hearted) attempts to mitigate the Global Warming problem, it continues, unabated. This great global change not only affects us all and the world to which we have become accustomed, but it affects the rest of the world's biota as well. Corals and coral reefs are not immune. The result is what we call "coral bleaching." a result of climate change occurring especially in Earth's mid-latitudes where corals are particularly abundant, and is particularly well known from Australia's Great Barrier Reef. What is going on? The coral's algal endosymbionts are responsible for the unique and beautiful colors of coral reefs. However, when corals become environmentally stressed, as has occurred due to the temperature increase of their surrounding waters, their polyps expel the enclosed algal cells and the colony takes on a ghostly stark white appearance. If this "bleaching" of the corals goes on too long without zooxanthellae becoming re-established, it will result in the coral's death. And that is what we are witnessing.

Virtually all eumetazoans are bilaterally symmetrical

In contrast to the radial symmetry of jellyfish and corals, all of the advanced animals you know – from termites to turtles, worms to walruses, butterflies to bats – are bilaterally symmetrical, their right sides like their left **(Fig. 5-6)**. Why is the right side of *your* body a mirror image of the left? The answer, of course, is evolution – all life is linked. But how and why did this particular and rather simple bilaterally symmetrical organization arise and why is it so widespread, so obviously successful?

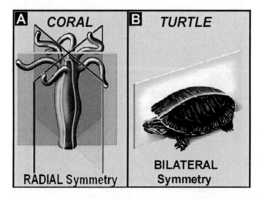

Fig. 5-6 (A) Jellyfish and corals (coelenterates) have radial symmetry whereas **(B)** more advanced bilaterians have bilateral symmetry.

The first link in this evolutionary chain is occupied by lowly flatworms, members of the Phylum **Platyhelminthes** (from Greek *platús*, "flat" plus *hélmins,* "parasitic worm," platyhelminthes thus meaning "flat parasitic worms"), a group also known as "planarians." Unlike radially symmetric coelenterates, flatworms are lengthwise bilaterally symmetrical, small worm-like creatures that use their cilia-coated undersurfaces to crawl about **(Fig. 5-7)**. Indeed, it is this act of crawling, of moving about on surfaces – rather than floating in the ocean like a jellyfish or being anchored in place like a coral – that explains their symmetry. To be an effective mobile surface dweller an animal has to move around to locate food and find a mate, movement that is facilitated and actually enabled by having a right side that is equal to the left coupled with a differentiated top (dorsal) and bottom (ventral) part of the body and a head-to-tail organization. Because this organizational pattern has proven so successful, it has been carried over to all later evolving members of the Animal Kingdom, all of which are thus included in the "Bilateria" **(Fig. 5-2)**.

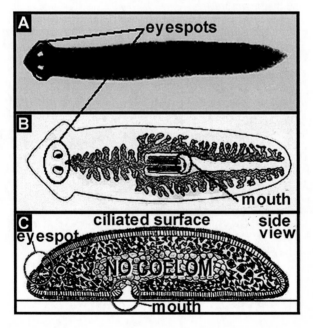

Fig. 5-7 (A-C) Flatworms (planarians) have eyespots, bilateral symmetry and ciliated surfaces which allow them to move across surfaces to vacuum up scattered food particles, but unlike more advanced animals their bodies **(C)** are solid, packed with cells, lacking an open space (the coelom) between the gut and the body wall.

Like all other bilaterian eumetazoans, flatworms have three layers of tissues, not just the ectoderm and endoderm, like those of jellyfish and corals, but an additional middle layer as well, the mesoderm. But unlike more advanced animals, the entire bodies of flatworms, extending from their centrally located tubular gut (intestine) to their outermost skin, are completely filled with cells (**Figs. 5-7** and **5-8**). At this point in the discussion you may mummer to yourself, "What's that about? We are advanced animals … aren't our bodies solid, completely packed with cells?" Not so! Think about it. If our bodies were packed with solid tissue there could be no such thing as open-heart surgery. Opening up a person's chest would not expose the heart, lungs and other vital organs. Instead we, like all like all other advanced animals, have an empty gap, a void between the central alimentary canal, the gut, and our outer body walls, an open space known as a **coelom** (from Greek, *koilōma*, "cavity"). Advanced animals all have coeloms, all are thus "coelomates," whereas flatworms are "*a*coelomates" (a term highlighted by use of the Greek prefix *a-,* "not," a usage like that in *a*moral).

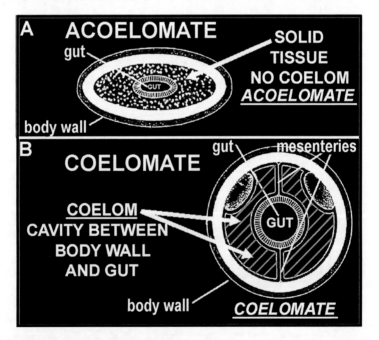

Fig. 5-8 (A) In acoelomates, such as flatworms, the gut is encased in solid tissue whereas **(B)** more advanced coelomate animals, such as humans, have a coelom, a cavity between the gut and the body wall.

Yes, acoelomate eumetazoan flatworms are small and primitive – "lowly" – but they are also terrifically interesting, primarily because they are the earliest evolved animals to exhibit a central nervous system that includes a well-organized brain housed at the front end of their bodies, the "encounter end" that first interacts with the environment as they move about. Spurred by the presence of this earliest evolved brain and by the flatworms' easy to discern forward-looking two eyespots, they have long been of great interest not only to biologists but to neurobiologists and psychologists as well, often used as an animal model in neurological research. Indeed, there are a great number of similarities between flatworms and humans, both exhibiting essentially the same organ systems – a brain and a central nervous system, eyes, musculature, intestine, epidermis, and distinctive reproductive structures. However, unlike humans and other more complex advanced animals – but of particular usefulness in human-related research – flatworms have an amazing capacity to re-grow, to regenerate parts of their bodies following their loss or injury. Why does that matter? Here's the answer.

Early experiments showed that flatworms shy away from flashes of bright light, an easily understandable response given their penchant for inhabiting lightless murky environs. Yet it is the flatworms' brain and its biochemically encoded memory bank that largely accounts for the great interest in these earliest evolved bilaterians. The seminal studies of flatworm memories date from the 1950's and 60's and the research of University of Michigan animal biologist and psychologist James V. McConnell (1925-1990). Best known for his investigations of learning and memory-transfer in flatworms – experiments recounted in psychology classes and texts worldwide – McConnell showed that flatworms could learn to navigate a Y-shaped maze in search of food and that if they were then decapitated, their heads severed from their bodies, they could regenerate their brains and head parts. Even more remarkably, he showed that if such rebuilt flatworms were fed the minced up remains of other flatworms that had previously learned to traverse the maze, they acquired the ability to unerringly navigate the very same maze!

Taken together, these experiments document a degree of regenerative plasticity and memory-transfer completely unknown among more advanced animals. Not surprisingly, McConnell's unexpected research results were immediately questioned and thoroughly scrutinized – some scientists accepting his reports verbatim but the great majority remaining highly skeptical. And although some aspects of his studies soon proved difficult to replicate they spawned a 15-year-long episode of serious scientific study of memory transfer, not only among flatworms but in rats, octopuses, praying mantes, baby chicks, kittens and honey bees, results reported in nearly 250 published scientific papers. Quite clearly, McConnell had showed that something truly unexpected was going on; and his results were encouraging enough that others wanted to find out for themselves.

Although the initial furor has long-since died down – an uproar exacerbated by McConnell's seemingly self-serving celebrity status-seeking persona – the scientific questions raised by his findings have not gone away. As recently as 2013, for example, T. Shomrat and M. Levin of Tufts University repeated the central aspects of McConnell's studies but this time, rather than using a maze they used simulated natural settings (comparing the flatworms' memories of a rough food-sprinkled surface and a smooth surface devoid of food). Further, to avoid the sort of "observer bias" of which McConnell had been accused, they used a computerized automated system to record the flatworms' movements. Afterward, all of the worms were decapitated and, as expected, within the next two weeks they regrew their heads and brains. Lo and behold, when the worms were put back onto both types of surface, the group that had previously associated the rough surface with food moved about and eagerly ferreted out the available fodder whereas the other group held back, hesitant to venture from their previously learned smooth surface setting.

Though the final chapters of this now 70-year-long saga have yet to be written, it is abundantly clear that small, primitive, acoelomate bilaterian eumetazoan flatworms – the earliest evolved animals having a central nervous system and a well-organized brain – are enormously interesting.

Why do coeloms matter?

All animals more advanced than flatworms have a gut-to-outer-wall empty space, a coelom, even a Thanksgiving Day turkey **(Fig. 5-9)** – which gives you an excuse to have some fun next Thanksgiving by pointing out to your mom that she is "stuffing that stuffing stuff, the 'trimmings,' into a *coelom!*" As we have seen repeatedly, any evolutionary innovation that is widespread in the descendants of a given group, whether in animals (for example, vertebrate limb bones; **Fig. 1-6**) or in plants (megaphyll leaves; **Fig. 3-3**), must initially have been of great advantage to its originating lineage in the never-ending battle to succeed and thrive. The coelom, present in all animals more advanced than acoelomate flatworms, is no exception, a true winner carried over to all descendants.

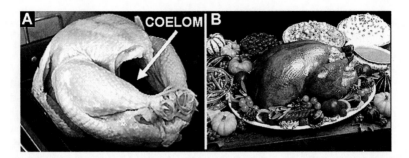

Fig. 5-9 All coelomate animals, such as a turkey **(A)**, have an empty cavity, the coelom, between their gut and outer wall. **(B)** Once the turkey innards have been removed, this leaves a large open space where vegetables and the other "trimmings" are commonly cooked for a Thanksgiving Day dinner.

No doubt this is true, but just why do coeloms matter, what indispensable function did they initially play that has locked them in to all later-evolving animals? Though the development of bilateral acoelomate flatworms was a major step in animal evolution, chiefly because these primitive cilia-coated planarians can move across surfaces to vacuum-up bits and pieces of food particles that have filtered down from the overlying waters, such surficial detritus represents only a small fraction of the food available in a sediment. Think about it. Day after day, year after year, microbes, plankton, seaweeds, fish, and all other marine organisms die and their remains sink into the abyss. Over time, their decaying remains become buried in silts and sands eroded from the land surface. Thus, the sediments that underlie the ocean bottom are chock-full of edible organic matter, a huge potential resource of food. But this underlying goldmine is unavailable to flatworms, simply because they can move only across surfaces whereas access to the fodder below would require them to burrow down into the mud to get at it. In essence, that is the why and wherefore of the presence of coeloms in all later evolved animals.

In coelomates, humans included, the gut is suspended in the coelomic body cavity by intermittently spaced strong connecting bands of tissue, "mesenteries" **(Fig. 5-8B)**, with the body wall serving to shield this centrally situated vital organ. This basic tube-within-a-tube organization rather closely resembles that of a flexible coaxial cable **(Fig. 5-10C)**, a type of electrical conduit in which an inner electricity-carrying wire is surrounded by a concentric protective shield, the term "coaxial" referring to the shared central axis of the conductor and the outer casing. Used commonly in telephone trunk-lines and broadband internet connections, it is this structure that gives such cables their required flexibility. Not surprisingly, it serves the same function in animals. Thus, and although in coelomates the body wall and the gut are linked by bands of tissue, each can move separately – the body wall "pushing" in one direction and the gut "pulling" in the other. And that back-and-forth movement, a "wriggling" motion that is easy to observe as an earthworm burrows its way into an underlying substrate **(Fig. 5-10B)**, is why the coelom was initially so important. With this innovation, coelomates could tap a rich new reservoir of food unavailable to their surface-limited acoelomate competitors, and this in turn allowed the newly evolved burrowing coelomates to succeed, to thrive, to "win" the competition. Thus, flexibility and the accompanying ability to burrow are the fundamental reasons why the coelom was carried forward to all later evolved animals.

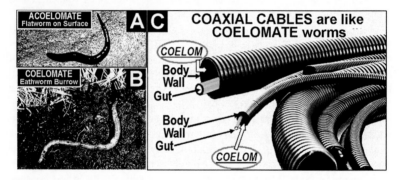

Fig. 5-10 Acoelomates, such as a flatworm **(A)**, can only move about on surfaces whereas coelomate animals, such as an earthworm **(B)**, can burrow, the open space in their bodies providing flexibility like that of **(C)** coaxial electrical cables.

CHAPTER 6

"SEA SHELLS" AND THEIR KIN

Protostome coelomates – the lineage leading to insects

As we have seen in **Chapter 5**, insects account for about 80% of all named species of animals and more than half of all the catalogued types of life. If you think about it, that seems rather odd. After all, we don't think of insects as mattering much at all. We notice them from time to time, yet except if they infest our homes (like termites or scurrying cockroaches) or occasionally sting us (like bees or wasps), they don't bother us. Moreover, if insects are all that numerous, why aren't we ankle-deep in their dead or dying carcasses? Love 'em or hate 'em, they ought to be a lot more obvious!

The answer is simple. Ground-dwelling Insects are foraging omnivores, "phagotrophs," able to eat virtually any organic remains they come across, even the dead bodies of other insects. Have a look at the ants in your backyard. They travel around on the lookout for food to carry back to their nest, each ant following the path of the ones before (led by the scent, the "pheromones" laid down by ants that already traveled the path). If they find a tasty morsel, say a dead dragonfly, they munch it up and haul back pieces to their nest. And if they come across a dead member of their own "tribe" – identified by the pungent formic acid smell it gives off – they will pick it up and add it to their ant colony's "funeral pile." Without doubt, insects are the land world's foremost "clean-up crew."

How did insects arise, what are their evolutionary roots? Like the great majority of non-backboned (more formally, "invertebrate") animals, insects are protostome coelomates, the subject of this chapter **(Fig. 6-1)**. The term coelomate we understand, but what is a "protostome"?

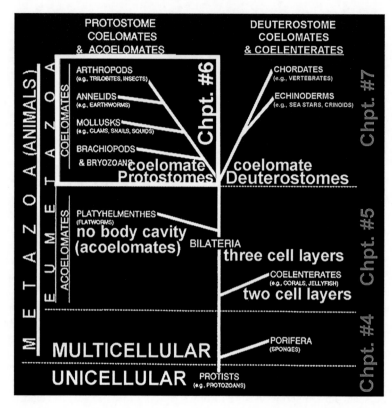

Fig. 6-1 A simplified phylogenetic "Tree of Life" of the Animal Kingdom showing in the white rectangle the part of the tree discussed in **Chapter 6**.

In coelomate animals, a sperm fertilizing an egg produces a two-celled body, the zygote, which then divides repeatedly to ultimately form a hollow ball termed a "blastula" **(Fig. 6-2)**. Over time as the blastula continues to grow, a specialized circular opening known as the blastopore then forms at one pole of the increasingly larger blastula and the wall encircling the blastopore in-folds ("invaginates") to produce a blastula-penetrating hollow tube. This tube then grows down the center of the blastula to the opposite pole where it penetrates the outer wall to produce a second circular opening. In all coelomate animals, this central tube becomes the alimentary canal, the "gut." In coelomate **protostomes** (from Greek *prot-*, "first," plus *stoma, "mouth"* – "first mouth") the first-formed hole, the blastopore, becomes the mouth whereas the later-formed opening becomes the anus.

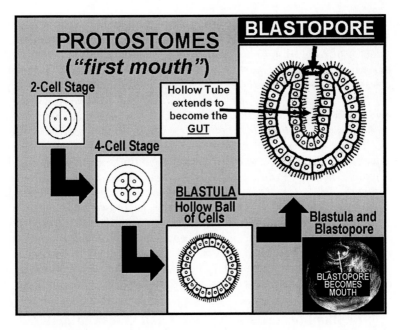

Fig. 6-2 In protostome animals, the hollow embryonic blastula in-folds at the blastopore to form the central tubular gut of the adult animal, the initially formed hole becoming the mouth and the exit pore becoming the anus.

During later development, the blastula stage of all protostome coelomates exhibits a so-called "holoblastic" spiral pattern of cleavage, an important difference from the radial cleavage pattern of the blastulas of more advanced deuterostome animals (to be discussed in **Chapter 7**). Using phylogenetic trees based primarily on 18S rRNA (ribosomal ribonucleic acid), the protostome invertebrates have been divided into two great sister-groups (called "clades"), the Spiralia (including bryozoans, brachiopods, mollusks and annelids) and the Ecdysozoa (housing arthropods and, thus, all crustaceans and insects, along with a few other groups). Although introduced to such ordering systems only rather recently, the Spiralia/Ecdysozoa system has already gained substantial acceptance among invertebrate zoologists, many of whom are enthusiastic adherents of genome-based phylogenetics. To date, however, a sizeable portion of the invertebrate paleontological community remains skeptical.

To those unacquainted with the workings of science, this current disparity between invertebrate zoologists and paleontologists – the community of scientists most knowledgeable about protostome invertebrate animals – may seem surprising. After all, they all study the same set of animals! In fact, however, the disparity is easily explained. Zoologists, students of extant life-forms, tend to put great stock in the genetics of modern organisms. In contrast, paleontologists rely primarily on the observable fossil record for which the preserved organisms, they (correctly) assert, *"had some genes that differed from those of their modern counterparts." "In addition,"* many would then note, *"the zoological phylogenists almost always rely on the fossil record to reveal the time of emergence of the various groups – their phylogenetic trees show the relationships among **living** organisms, **not** their ancient ancestors, **not** their extinct relatives, trees that are incapable of accurately dating the times of origin and divergence of the various lineages."* Though this debate may not be fully resolved for a decade or two, it will be useful to you to have here become at least a bit acquainted with the proposed Spiralia/Ecdysozoa concept, simply because it might eventually turn out to be the universally accepted accurate description of reality, the common goal of all scientists struggling to unravel the history and relationships among the numerous groups of protostome animals.

Lophophorates – "arm foots" and "moss animals"

Though the basic concepts underlying protostome coelomate evolution are not difficult to fathom, their easy understanding is plagued by unfamiliar scientific terms for which too often there are no commonly used words that convey the proper meaning. For example, a great many of such marine protostomes are what most of us would refer to as "sea shells," but that doesn't turn the trick – simply because there are myriad types of sea shells and what is of interest to us is how those many interrelated types arose over geological time.

Unfortunately, this "new terms" problem is all too common in science, shown here by three scientific terms that will be new to many, namely "lophophore," "brachiopod," and "bryozoan." For these, as for most scientific terms, it helps to know their linguistic derivation. Most are derived from ancient Latin or Greek, so-called "archaic languages" that are no longer taught in most high schools (though for me, among all the material to which I was there exposed my 10th grade Latin course has been probably the most useful). All three of the new scientific terms are derived from Greek: **lophophore**, from the Greek *lophos*, "tuft" plus *-phoros*, "bearing" – "tuft bearing"; **brachiopod**, from the Greek *brakhion*, "arm," plus *-pod*, "foot" – "arm foot"; and **bryozoan**, derived from the Greek *bryon*, "moss," plus *zôion*, "animal" – "moss animal."

Brachiopods **(Fig. 6-3A** to **C)** and bryozoans **(Fig. 6-3D** to **F)** are both situated close to the base of the protostome coelomate part of the phylogenic tree **(Fig. 6-1)**, grouped together because unlike other members of the lineage leading to insects they are lophophorates, both having leaf rake-like lophophores used to scoop-up food particles from their immediate environment. Interestingly, however – and although brachiopods and bryozoans are also similar by having their living tissues protected in rock-like calcitic ($CaCO_3$) encasements – the two groups markedly differ. Brachiopods, "arm foots," are relatively large easy-to-see seashells, anchored in place by a fleshly "pedicle," the arm-foot that protrudes from a hole in the shell into the underlying substrate. Their leaf-rake lophophores are also large, filling the interiors of their shells. In contrast, individual bryozoans are tiny, arranged side-by-side, cheek-to-jowl in fan-shaped and even reef-forming colonies **(Fig. 6-3F)**. As implied by the name "moss-animals," the upper surface of a bryozoan colony has a mossy appearance, a soft fluffy look resulting from row after row of lacy lophophores extending out into the surrounding waters.

Of the two types of lophophorates, brachiopods are probably the more familiar, primarily because they were particularly abundant in near-shore sea-bottom settings throughout much of the early Paleozoic, only later to be supplanted by clams, two-shelled (bivalved) mollusks that unlike sessile brachiopods can move, albeit rarely and slowly, across the sea-floor bottom with some mollusks, such as scallops, able to swim by clapping their shells together.

Moreover, we also know brachiopods from their use on the institutional logos of numerous universities, there depicted by a symbol of excellence, the "lamp of learning" and the reason why brachiopods are sometimes referred to as "lampshells." Once again, this may seem odd. What is lamp-like about a simple seashell? In fact, however, such use has a long historical basis. Indeed, this widely used symbol dates from at least as early as the time of King David (about 1,000 BC) in the Holy Land (Israel-Palestine), some 3,000 years ago, when brachiopod shells were used as small lamps, the shell half-filled with oil and the "arm-foot" pedicle replaced by a lighted wick. The shells first used were those of *Terebratula* and *Lingula,* receptacles later replaced by clay facsimiles.

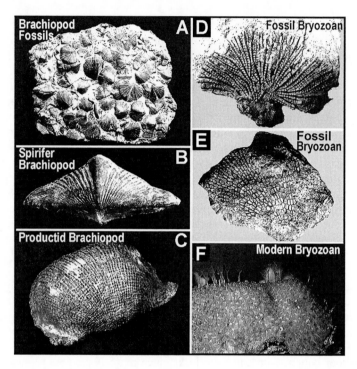

Fig. 6-3 (A-C) Early Paleozoic brachiopods and fossil **(D, E)** and modern bryozoans **(F)**.

Mollusks – "axe-foots," "stomach-foots," and "head-foots"

As we continue our march up the protostome coelomate lineage toward insects, we are once again beset by new terms, in particular, "mollusk," "pelecypod," "gastropod" and "cephalopod." The first of these, mollusk (sometimes spelled "mollusc") – the informal name for the Phylum **Mollusca** -- is the all-embracing term for the other three newly introduced terms, each referring to a particular type of invertebrate animal (as in "pelecypod mollusk," "gastropod mollusk," and "cephalopod mollusk"). Surprisingly, the linguistic derivation of Phylum name Mollusca seems out of place since the term comes from the Latin word *mollis*, meaning "soft," an epithet that clearly does not fit given that most types of mollusks have hard shells, their bodies enclosed in the calcium carbonate hull secreted by their flesh-covering soft mantle. Instead, however, the name refers to the soft meaty texture of mollusk bodies, whether they are encased in a shell (like clams or snails) or are shell-less (like a land slug or an octopus).

Among all types of modern invertebrate animals, mollusks make up the second most diverse major group (outnumbered only by the insect-containing arthropod phylum), represented by more than 85,000 living species and a comparable 60,000 to 100,000 species of fossils. Moreover, by comprising nearly a quarter of all named marine organisms they are the largest phylum of formally catalogued marine life-forms. Taken together, mollusks include a wide range of animals familiar to us – mussels, scallops and oysters; limpets, snails and slugs; and squids, cuttlefish, and octopuses. Of the eight or nine major groups (taxonomic classes) of mollusks currently recognized (including two now extinct), only the three most prominent in the fossil record are here considered **(Fig. 6-4)**: pelecypods (scallop-like bivalve mollusks), gastropods (snail-like mollusks), and cephalopods (nautiluses and octopus-like mollusks).

Pelecypods (from Greek *péleky(s)*, "axe," "hatchet" plus Latin -*poda*, "foot" – "axe-foot"), so named based on the axe-like shape of their extended foot **(Fig. 6-4A)**, are also known as the Bivalvia, "bivalve mollusks," in reference to their two shells, the hinged together bilaterally symmetrical calcium carbonate "valves" that enclose their bodies. Of all types of pelecypods, the various sorts of clams – particularly the edible clams, the cockles and the Venus clams – are by far the most diverse, in total numbering more than 5,000 living species. Unsurprisingly, given their abundance and the relative ease of removing their encasing shells, the soft fleshy bodies of such bivalves have been harvested as a human food-source for thousands of years.

Most clams are marine and sedentary, protecting themselves from predators such as sea stars by burying themselves in seafloor sediment, a mode of life that promotes their abundant preservation in the fossil record. But some, such as the common scallop (various species of the genus *Pecten*) can plunge themselves about, surging through the water in fits and starts by repeatedly clapping their shells together, rapidly traversing short distances and occasionally even migrating to a new habitat. One such example of a fossil scallop is *Chesapecten jeffersonius*, particularly well known as the official state fossil of Virginia and named in honor of U.S. President Thomas Jefferson (the founder of vertebrate paleontology in North America). Finally, though most pelecypods are at most a few inches across, the world's largest should be noted – *Tridacna gigas* – the giant clam of the South Pacific (an endangered species no longer hunted but that in days past only a single specimen of which could provide a feast for multiple families).

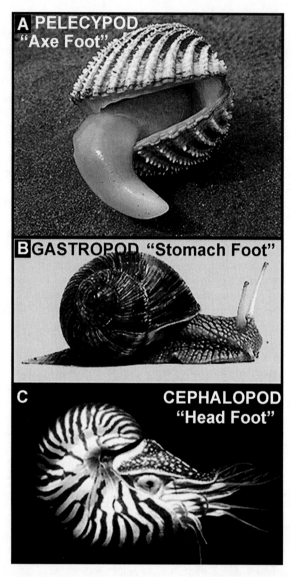

Fig. 6-4 Living examples of the three types of mollusks discussed in the text, **(A)** a pelecypod, a bivalve mollusk, showing its extended "axe-foot"; **(B)** a gastropod, a snail, and its "stomach foot"; and **(C)** a cephalopod, the chambered nautilus, with its tentacle-laden and eye-containing "head foot."

The second group of mollusks considered here are the **gastropods** (from Greek *gastro*, "stomach" plus Latin *-poda*, "foot" – "stomach-foot"), a moniker that refers to the large fleshy belly on which they slowly glide about **(Fig. 6-4B)** and a group that includes such invertebrates as snails, whelks and slugs. Although there are more than 65,000 described species of gastropods – making them the largest of all mollusk groups – to most of us they pass unobserved, primarily because their land-inhabiting members tend to stay hidden during the day and are active only at night. Still, if you are a home-gardener, each morning you know all too well that they have visited your plot, chomping the leaves of your carefully cultured tomatoes, lettuce and celery and leaving behind their drying slimy trails.

Most members of the group, the snails and whelks for example, have a single coiled ("spired") topknot shell into which the animal can withdraw, their bodies protected at the bottom of the shell by a trap door-like operculum. In contrast, land slugs are snails in which the enclosing shell has been reduced to an internal fragment or completely lost over the course of evolution. Interestingly, the local the abundance of snails and slugs can be enormous, reportedly millions of brackish-water and freshwater species cohabiting small mud flats; a single acre of British farmland said to hold 250,000 slugs; and a Panamanian mountain forest being estimated to have 7,500,000 land snails per acre. Evidently, gastropods are all around us, even though we barely notice!

The third type of mollusk to be discussed here are **cephalopods** (from the Greek *kephalo*, "head" plus Latin *-poda*, "foot" – "head-foot"), an epithet derived from the prominent head-encircling tentacles that they use to feed and move **(Fig. 6-4C)**. The basic body plan of most cephalopods includes two eyes, at least eight arms, a "mantle" – the fleshy body layer that secretes their protective shell – and a tube-like siphon that connects to the outside at the head area and runs down the interior of the shell, a conduit used to pipe water in and out and thus control buoyancy, the animal's up-down position in the water column. Some have hard cartilage-like internal structures – for example, the "cuttlebone" in cuttlefish and the "pen" (known also as the gladius) in squid – that evolved from the calcitic outer shell of their ancestors but that in most varieties of octopuses has been lost over evolution.

Of all modern cephalopods, only the chambered nautilus **(Fig. 6-4C)** has an external shell. This, however, was not the case in the geological past and, thus, in contrast to familiar soft-bodied cephalopods such as octopuses, cuttlefish, squids and the like, ancient shelled cephalopods have an excellent fossil record. The best known and most prominent of these are fossils known as "ammonoids," a now extinct lineage dating from the

Devonian to the end of the Cretaceous and particularly abundant in marine Mesozoic sediments that like much of Earth's biota perished during the Cretaceous-Tertiary great extinction event **(Chapter 3)**. The shells of cephalopods, like those of modern nautilus, may well have exhibited what is know as "counter-shading," being light-colored on the bottom and dark on top, a color pattern that by blending in to the surrounding waters, whether viewed from above or below, helps to avoid predators. Interestingly, Pliny the Elder, a Roman naturalist and natural philosopher who died in 79 A.D. during the volcanic eruption of Mt. Vesuvius that overwhelmed Pompeii, was the first to use the name "ammonoid." In his writings, Pliny referred to the fossils as *ammonis cornua*, meaning "horns of Ammon," a name he coined to honor the Egyptian god Ammon *(Amun)* who was typically depicted wearing curled ram's horns that have a shape similar to the spiral shape of most ammonite fossils.

Although some fossil cephalopods, those commonly known as "nautiloids" had large, straight, torpedo-shaped shells – as much as 13 feet (4 m) in length – the great majority had flat spiral shells like that of the living chambered nautilus, *Nautilus pompilius,* also called the "pearly nautilus." The interiors of such shelled cephalopods are subdivided into a series of chambers by calcitic cross-walls called "septa," chambers that became increasingly larger as the animal grew, the front-most chamber, the most recently formed in the sequence, housing their fleshy body.

As paleontologists have repeatedly shown, the range of shapes of the easily discerned linear markings on the outer surface of ammonite shells that indicate where the internal septa intersect the shells, markings known as "suture lines," systematically increased over geological time as the diversity of ammonites gradually expanded. Two near-endpoints of the sequence well illustrate this evolutionary change, beginning with the simple wavy to V-shaped markings of early-evolved (Carboniferous-age, 350 Ma) *Goniatites* **(Fig. 6-5A)** and proceeding to more convoluted sutures such as those of the Cretaceous-age (100 Ma) ammonite *Placenticeras* **(Fig. 6-5B)**. Because the internal mineralic septa support and provide strength to the encasing outer shell, the suture lines indicating the shape and therefore the overall length of the septa, shells having convoluted sutures are appreciably stronger than shells in which the sutures are simpler. Thus, the addition of ammonites having increasingly complex sutures evidently reflects a change in the environments in which they could live, the early-evolved species mostly inhabiting relatively near-surface waters and the gradually added new species able to withstand the higher hydrostatic pressures of increasingly deeper waters.

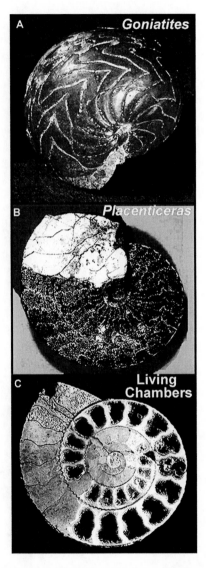

Fig. 6-5 Cephalopod ("head-foot") ammonite mollusks, **(A)** *Goniatites*, a Carboniferous (325 Ma) ammonite, showing the simple suture lines on the surface of its shell; **(B)** *Placenticeras*, a Cretaceous (100 Ma) ammonite exhibiting convoluted sutures; and **(C)** a sliced Jurassic (175 Ma) specimen illustrating its living chambers partitioned by internal septa.

Ammonites have also played a prominent role in folklore, most notably in Ireland and England, even helping to explain the roots of St. Patrick's Day (the yearly, March 17 "wearing of the green"). St. Patrick, commonly depicted in green vestments and the primary patron saint of Ireland, was the fifth-century missionary credited with having converted the Celtic polytheistic Irish society to Christianity. According to legend, Patrick then banished all of Ireland's snakes, chasing them into the sea after they attacked him while he was secluded during one his Biblical 40-day fasts. Where did the exiled snakes end up? Predictably, as the story has it, they invaded England – Ireland's larger, more powerful nearby nemesis. But St. Hilda of Whitby (614-680), an early Christian saint and the founding abbess of the monastery at Whitby in northeast England, would have none of this. On a wind-swept coastline near Whitby plagued by these writhing unwelcome intruders, St. Hilda, with eyes closed and head bowed, prayed for God's power to exterminate the snakes, traditional symbols of evil in Christian lore stemming from Eve's transgression in the Garden of Eden. Her plea was answered – not only did her prayer turn every snake into stone but it also decapitated them in the process – evidenced by the headless ammonite fossils common in the nearby Jurassic sediments. During the Victorian Era (1837-1901) this gave rise to a local cottage industry that produced St. Hilda's "snake stones" replete with their restored heads **(Fig. 6-6).**

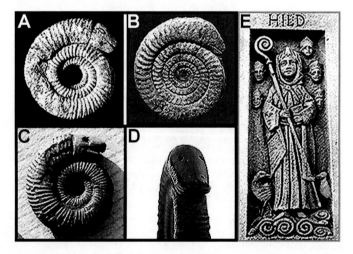

Fig. 6-6 "Snake stones" **(A-D)**, Jurassic (175 Ma) ammonites adorned with carved heads, and **(E)** a monument to St. Hilda, "HILD," at the monastery she founded at Whitby in northeast England, showing below her feet the vanquished snakes.

Segmented worms with legs link annelids to arthropods

The next group in line in the unfolding saga of protostome coelomate phylogeny are the **Annelida**, "annelids" (from Latin *anellus*, "little ring"), known also as "ringed worms" or, more commonly, segmented worms. Characterized by their regularly segmented body plan, as their name implies, and dating from the Cambrian Explosion of Life, well-preserved fossil annelids are easily identified. Like the Mollusca, the Annelida is a large phylum containing over 22,000 living species including such familiar creepy-crawlies as leeches and common earthworms.

It is true enough that most of us don't care much about annelids. Yes, in the Middle Ages and again in the 1800's, bloodsucking leeches were used widely in medicine to offset inflammation, and many of us know that earthworms, by burrowing into the soil, aerate gardens and grassy lawns. Other than that, annelids are not of much interest. In my youth, however, I was curious about them. Here's the snippet tale. When I was in the 3rd or 4th grade, my older brother would take me off to go fishing at a nearby river. My job was to dig up the bait, earthworms, from the backyard. The bait worked. I wondered why? Maybe they taste good? I captured one, stuck it in my mouth, and sliced it with my teeth. No taste at all, just a slimy wall filled with grit! Immediately I spit it out. (Thankfully, now that I've done that test, you don't have to – which I do not recommend.) Still, this little experiment answered my quandary. Earthworms are good fish-bait, not because they taste good but because ensnared on a hook they wriggle about, their movement attracting fish to chomp them down, and an annelid's grit-filled body results from the way it feeds, tunneling through the soil and then extracting in its gut whatever bits and pieces of food the in-taken dirt may contain.

Has the annelids' regular segmentation carried over to later-evolved lineages? The answer is a resounding "yes," well evidenced in arthropods, the immediately following major lineage of the protostome sequence and shown even by the regularly segmented structure of the backbones of all vertebrate animals including humans. The underlying genetic bases of such regular segmentation – an obviously highly successful innovation – are known as "Hox genes."

What are Hox genes and how do they work? In essence, they are a chain of related genes encoded in the DNA of chromosomes that specify regions of the body-plan of an animal embryo along its central head-to-tail axis, the protein molecules there defined ensuring that the correct structures form in the correct places of the body. Thus, annelids, arthropods and vertebrates have similar Hox genes, in each case clustered together in discrete locations on a chromosome with their ordered arrangement reflecting an anterior to posterior order of their expression. For example, Hox genes in protostome coelomate insects specify the order of appearance of the various segments and of the appendages that then form on a given segment (for example, legs, antennae, and wings) whereas the Hox genes in vertebrates specify the order, types and shapes of the backbone vertebrae that will form. The precise evolutionary origin of Hox genes is difficult to pinpoint, but their marked similarity one to another in a given sequence suggests that they are a product of gene duplication, the founding gene in the original sequence having been repeatedly copied and the whole assembly then being carried over and modified in later-emerging evolutionary lineages.

Although the precursors of Hox genes may be present in sponges and no doubt occur in mollusks and other protostome coelomates, their presence is particularly well evidenced by wormlike animals known from the Cambrian Explosion of Life. A good example is the modern peanut worm *Dendrostomum* and its fossil equivalent, *Ottoia*, earliest known from Walcott's Burgess Shale Fauna **(Fig. 6-7)**.

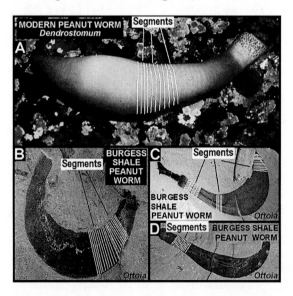

Fig. 6-7 The existence of Hox genes, evidenced by the occurrence of segmented wormlike animals such as (**A**) the modern peanut worm *Dendrostomum,* date from at least as early as the Cambrian Explosion of Life as shown by **(B-D)** three specimens of the Burgess Shale 510 Ma peanut worm-like fossil *Ottoia.*

Such soft-bodied vermiform animals differ significantly from the next great group in the evolutionary sequence, arthropods, animals in which the segmented body exhibits prominent protruding jointed legs. Fortunately, we have what was once a missing link, a so-called "living fossil" to fill in this gap **(Fig. 6-8)**, the modern segmented wormlike animal *Peripatus* that has pair after pair of short annulated stubby legs – a worm with legs! First described in 1825, *Peripatus* was the earliest named example of the Phylum **Onychophora** (from Greek *onyches*, "claws," plus *pherein*, "to carry" – "claw-carrier"), a small taxonomic group more commonly known as "velvet worms" due to their velvety texture and wormlike appearance. The bodies of such worms, generally only a few inches in length, are covered by a thin veneer of skin-like chitin enshrouded by the rows of water-repellent overlapping scales that produce the worms' velvety appearance. Given these traits and the velvet worms' parallel rows of paired annulated legs they inhabit damp tropical forests where at night they actively crawl about, foraging for small insects among the scattered leaf litter, under stones or fallen logs, and even inside termite nests.

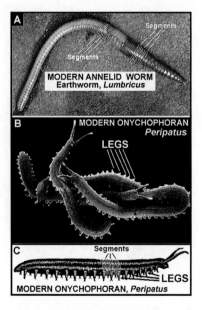

Fig. 6-8 (A) A modern annelid, the common earthworm *Lumbricus*, compared **(B, C)** with the onychophoran velvet worm *Peripatus* that exhibits a segmented body and annulated legs, an intermediate stage between annelid worms and arthropods.

Thus, unlike earlier evolved legless segmented worms, velvet worms are not subsurface-burrowers and they also differ by having high-powered "slime canons" that they use to squirt out volleys of sticky slime to as far as two feet away that surprise and ensnare their prey. Nevertheless, like a great many other groups of animals, velvet worms date from the Cambrian Explosion of Life. Indeed, as we have now seen repeatedly, the relatively short-duration, less than 60-Ma-long (541-485 Ma) Cambrian Period of geological time and the new animal lineages that then emerged marked a huge advance in the evolutionary development of virtually all modern groups of animal-life, the sole exception being bryozoan "moss animals" that are first known from the earliest segment of the immediately overlying Ordovician Geological Period.

Jointed-legged arthropod date from the Cambrian Explosion

Arthropods, relatively large, segmented, jointed legged animals, are familiar to us all. We eat them – shrimp, lobsters, crabs and the like (**Fig. 6-9A** and **B**) – and we watch crabs and similar arthropod crustaceans (from Latin *crūsta* "shell" or "crust' – "hard-shelled") scurry around on the beach. Moreover, because they have a preservable robust exoskeleton, their carapace, they have an excellent fossil record, arthropod trilobites being the "favorite fossil" of a great many amateur collectors.

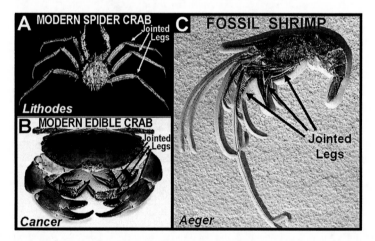

Fig. 6-9 Two modern jointed-legged arthropod crustaceans, **(A)** the spider crab *Lithodes* and **(B)** the edible crab *Cancer*, and **(C)** a well-preserved fossil crustacean, the shrimp *Aeger*, from the Jurassic (150 Ma) Solnhofen Limestone of southern Germany.

The name of the entire group, the Phylum **Arthropoda** (from the Geek *árthron*, "joint'" or "jointed," plus the Latin -*poda*, "foot" – "jointed -foot", "jointed-leg") highlights their appendages, their defining onychophoran-derived add-on to the body plan of legless segmented annelids. Like mollusks (and to a lesser extent, annelids), arthropods comprise a huge taxonomic group. In fact, the Arthropoda – numbering an estimated 7 million living species, including some 1.5 million types of beetles and 5.5 million species of insects – is the largest phylum of the Animal Kingdom. All members of the phylum have an exoskeletal carapace, a segmented body, and paired jointed appendages. Their exoskeletons are made of cuticle, a non-cellular organic material secreted by their skin, their epidermis, and each body segment and limb section is encased in hardened cuticle which in most aquatic crustaceans is further strengthened by being biomineralized with calcium carbonate.

Depending on the local environment, arthropod fossils can be exquisitely preserved. Among such deposits, one of the best known is the upper Jurassic (155 Ma) Solnhofen Limestone of Bavaria, in southern Germany, an exceptionally fine-grained calcitic rock unit in which numerous fossil animals are preserved in life-like detail including the front-end "feelers" of fossil shrimp **(Fig. 6-9C)** and even detailed imprints of the feathers of *Archaeopteryx,* the earliest known fossil bird. A second such *Lagerstätten* (a German term referring to a sedimentary deposit that exhibits extraordinary fossils having exceptional preservation, sometimes including preserved soft tissues) is the middle Cambrian (510 Ma) Burgess Shale of British Columbia, Canada, famous not only for providing the first insight into the great burst of Cambrian evolution but the remarkable preservation of its life-like fossils. Thus, as is true of annelids, velvet worms and most other types of invertebrate animals, there is firm evidence that arthropods date from the Cambrian Explosion of Life.

Because living and fossil arthropods are so very numerous, we know quite a lot about them. All have a life-history that goes through a series of marked sequential changes, a process formally known as "ecdysis" – in more common parlance, molting – the individual stages of the sequence known as "instars."

Why does this seemingly odd, charge-and-change-again molting process occur? Once again, the answer is easy to understand. All arthropods are enclosed in a stiff robust carapace that because it cannot stretch and expand as the animal grows, restricts growth. Thus, by the process of molting, arthropods shed their old exoskeletons, sequentially replacing them from one instar to the next with a larger roomier version. Molting, however, also has a downside. When first put in place the new carapace has not yet hardened, and because the arthropods are therefore unprotected and essentially immobilized until the new cuticle becomes robust they are in danger of being attacked by predators, a major cause of all arthropod deaths. Soft-shelled crabs provide a case in point, crabs that are harvested during their molting season so that restaurant diners do not have to bother with their enclosing stiff shells. Similarly, such molting explains why trilobite fossils are relatively common in Paleozoic sediments, the fossil finds being of cast-off exoskeletons not of the preserved whole animal.

Even though molting occurs only occasionally, molting cycles run nearly continuously until an arthropod reaches its full size, and most add layers to the inside of their exoskeletons virtually all the time, a daily growth increment that in some insects forms growth-rings similar to those of corals and trees. In addition, molting permits what is called "metamorphosis," the sequential addition of new body segments, new appendages, and even new external eye lenses, a process that accounts for the sometimes marked differences in the morphology and life-styles of successive instars such as those of butterflies and moths in which there are four distinct stages: egg, larva (caterpillar), pupa, and adult. Although the impetus for molting varies from one taxon to the next, it can be influenced by such external factors as temperature or day-length, as is illustrated by dragonflies that have broad geographic distributions. In one such species (*Tetragoneuria cynosure)* that goes through twelve instars from egg to larva to nymph to adult and ranges from Michigan to South Carolina, the entire process occupies two years in the northern reaches of its range but in the Carolinas happens twice as fast, occurring during a single year.

As an example of the many well-studied ancient arthropods known to science, let's now consider trilobites, many fossil-lovers "favorite fossil." Like annelids and velvet worms, trilobites also date from the Cambrian Explosion of Life **(Fig. 6-10)**. And even though they lived only during the Paleozoic Era (from the beginning of the Cambrian through the end of the Permian, spanning the time from 541 to 250 Ma), the trilobite life-history, their instar-recorded "ontogeny," is well documented in the fossil record **(Fig. 6-11)**. However, and despite such detailed knowledge and the relatively common occurrence of trilobites throughout the Paleozoic, a major question about their life-history has long remained unanswered. Namely, how did trilobites reproduce?

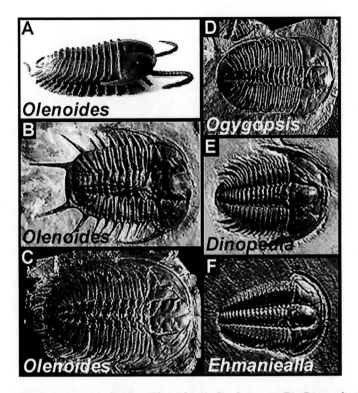

Fig. 6-10 The arthropod trilobites *Olenoides* **(A-C)**, *Ogygopis* **(D)**, *Dinopedia* **(E)**, and *Ehmaniealla* **(F)** from the 510 Ma Burgess Shale and the Cambrian Explosion of Life.

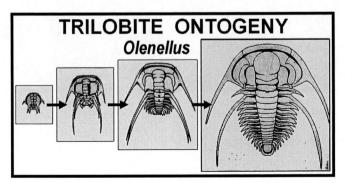

Fig. 6-11 The Paleozoic trilobite *Olenellus gilberti* showing fossil-based examples of the instars of its life history, its ontogeny.

To answer this question, we turn to the closest living relative of trilobites, *Limulus*, the modern horseshoe crab (early-evolved arthropods that have survived from the late Ordovician to the present) and ask how it reproduces. The answer is easy to figure out – *Limulus* reproduces by releasing clusters of eggs nestled under the frontal parts of its body-covering carapace **(Fig. 6-12)**. Might this also be true of long-extinct trilobites? Yes, indeed, as shown by T.A. Hegna (Western Illinois University) and his colleagues, M.J. Martin and S.A.F. Darroch (Vanderbilt University) who in 2017 reported discovery of a well-preserved specimen of the Ordovician (450 Ma) trilobite *Triarthrus* that exhibits similar egg clusters housed under the frontal parts of the fossil's carapace **(Fig. 6-13)**. Thus, trilobites reproduced by the same processes as those of their closest living relative. The question of how trilobites reproduced has finally been laid to rest.

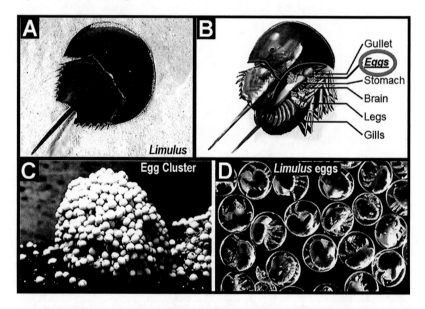

Fig. 6-12 (A) The horseshoe crab, *Limulus*, reproduces by **(B)** producing eggs that are housed below the frontal area of its carapace and **(C, D)** are released to the environment in large clusters.

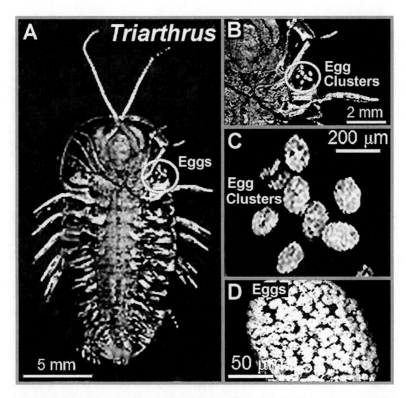

Fig. 6-13 (A, B) The Ordovician (450 Ma) trilobite *Triarthrus* reproduced by producing eggs below the front-part of its carapace, egg clusters **(C, D)** released into the environment in ovoid masses.

Yet another question comes to the fore, namely how did trilobites see, how did they avoid predators and ferret out food across the seafloor? The eyes of trilobites are bulbous **(Fig. 6-14)**, a lot larger relative to their head-size than those of most animals familiar to us. Moreover, innumerable circular to more commonly hexagonal facets, small calcitic lenses quite unlike the structure of our gelatinous smooth-surfaced eyeballs, pockmark the outer surface of their globose eyes. Why is there such a difference?

Fig. 6-14 (A-C) Like all other arthropods, the mid-Paleozoic (late Ordovician to end Devonian, 444-359 Ma) trilobite *Phacops* had **(B-D)** bulbous multifaceted compound eyes.

Like the solution to how trilobites reproduced, let's compare the fossils to their living relatives. Trilobites had compound eyes, much like those of other arthropods such as insects (**Fig. 6-15A** and **B**). Like human eyes, such eyes detect color and movement – as of course they must for their bearer to avoid predators. (You can check this out yourself by trying to catch a buzzing fly as it bothers you at dinner – it'll see you coming to smack it down! Indeed, the only person I've ever seen to truly master this feat was the World Boxing Champion Mohammed Ali whose hands were lightening fast.) And though the vision provided the multifaceted compound eyes of arthropods is not nearly as sharp as we humans take for granted, their bulbous form and relatively large size give them an appreciably larger, left-to-right and up-to-down field of view than we can see (**Fig. 6-15C**).

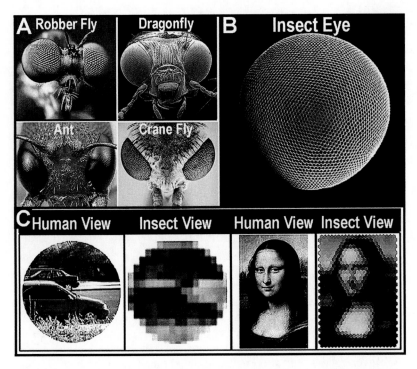

Fig. 6-15 (A, B) Insects, like trilobites and other arthropods, have compound eyes that **(C)** have a wide field of vision, detect color and movement but yield a speculated blurred view of shapes.

Here, before we move on to consider insects, the final group of the protostome coelomates to be discussed, it is useful to note that the last two questions – about how trilobites reproduce and how they see – illustrate an underlying basic tenant of the way knowledge of the history of life advances. Such progress comes from understanding the problems posed, learning how they are solved by modern analogues, and examining relevant fossils to determine whether ancient life tackled the problems in the same or similar way. It is no doubt correct as Darwin's friend and mentor Charles Lyell taught us in the early 1800's that *"the present is the key to the past."* But the reverse also holds true, that the past is the key to the present. In other words, as scientists look at the products of a process, whether modern or ancient, they ask the telling question *"Why are those products the way they are?"* For life on Earth – again, whether modern or ancient – evolution holds the answer. Students of the history of life do this all the time and it is plenty good fun!

Though terrifically diverse, insects have a lousy fossil record!

Of the several million living insect species estimated to exist, about 900,000 have been formally described, in total amounting to nearly 60% of all named species of extant organisms. Most of these belong one of seven main insect groups: beetles, bees, bugs, flies, butterflies, crickets, and dragonflies. Of this mix, beetles make up about 50%. Most experts attribute the great success of insects to their possession of a protective exoskeleton, their small size and, for many, their ability to fly – the last two traits permitting them to escape predators and disperse to new environments. In addition, virtually all insects can produce large numbers of offspring relatively quickly. In short, insects have mastered the two great requirements needed by successful lineages, to stay alive and to reproduce their kind.

The formally preferred definition of members of the Phylum **Insecta** is restricted to "small arthropod animals that have six legs and generally one or two pairs of wings." For our discussion, however, we will use the broadened, more commonly applied informal definition of insects, namely, "any small invertebrate animal, especially one with several pairs of legs." Indeed, that is what most of us mean when we refer to insects, its usage having the advantage of including not only members of the official seven categories but also such additional small arthropods as spiders, centipedes, ticks and the like, some of which have many pairs of jointed legs.

Regardless of how one defines insects, all would agree that given their dominant abundance in the living world, their known fossil record is paltry, deficient, simply lousy. In large measure, this is an understandable result of their makeup and the way they live. All insects lack hard mineralized parts – which immeasurably decreases the likelihood of their remnants being preserved – and, as noted at the outset of this chapter, ground-dwelling Insects are foraging omnivores, phagotrophs that feed on virtually any organic remains they come across including the dead bodies of other insects. As a rule, the odds of any particular individual ever becoming fossilized are minuscule with the odds being not much better for even an entire successful species unless it lives in large populations. For insects, however, such rules do not apply. As the foremost "clean-up crew" of the world's land surface, insects destroy their own potential fossil record before it has a chance to be preserved! Thus, and although fossil insects are known from a few *Lagerstätten* – the Jurassic Solnhofen

Limestone, for example, and rare deposits of Paleogene and Neogene tree-secreted tar-like amber – their fossil record is essentially unknown.

Not only are species of insects remarkably numerous, encompassing a great range of differing types, but they also vary greatly in weight, the smallest weighing less than 0.000,035 ounces (25 micrograms) and the largest weighing some 2.5 ounces (70 grams), a weight for the largest more than a million times greater than the smallest! Among the most awe-inspiring traits of insects, however, is their amazing strength. Ants, for example, are super-strong, able to lift and carry to their nest large chunks of insect body parts, entire whole insects, and even large leaves with another ant perched on top **(Fig. 6-16)**. For ants, such amazing feats are no doubt aided by the self-cleaning sticky pads on their feet which help them not to be overbalanced by their load, even when they are suspended upside-down. In fact, most ants have the ability to carry between 10 and 50 times their own body weight, the amount varying with the particular species. The Asian weaver ant, for example, can lift 100 times its own mass.

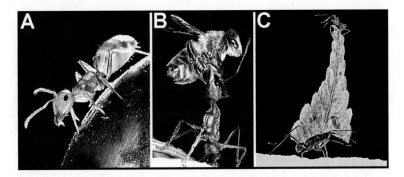

Fig. 6-16 Ants, members of the insect family Formicidae. **(A)** A representative specimen, **(B)** an ant hoisting a bee above its head and **(C)** a New World tropical leafcutter ant (a species of the genus *Atta*) hauling off a large leaf with another ant hanging-on at its top.

Pause for a moment and compare the strength of an ant with that of a human. The "World's Strongest Man," a much-coveted title, has recently been held by Lasha Talakhadze, a super heavyweight from Georgia – a small country on the border of Eastern Europe and Western Asia perhaps best known for its capital city, Tbilisi – who weighs in at 372 pounds (169 kg). In a recent Olympic competition, Talakhadze won the world weightlifting championship by breaking his own world record, amassing 1,067 pounds (480 kg) in the snatch and clean and jerk. In other words, he lifted a total of nearly four times his body weight – remarkable for us humans but a whole lot less relative to body-weight than feats accomplished day-in, day-out by members of your own run-of-the-mill backyard ant colony!

In addition to the seemingly super-human weight-lifting capabilities of ants, other insects and even their products are also super-strong, spider silk, for example, the thin glossy strands that spiders secrete to spin their webs are about five times as strong as the same weight of steel. And some insects are terrifically resistant to being crushed, even by a weight that for them is enormous. A case in point is *Phloeodes diabolicus*, a beetle more commonly known by its fearsomely unlikely moniker, the "diabolical ironclad beetle." Relatively common in arid areas along the western coast of the United States, it feasts on the fungi that live under the bark of oak and other trees. Like other beetles, during daylight it mostly stays out of sight and plays dead when it senses danger. Remarkably – and though diabolical ironclad beetles are tiny, not much bigger than a grain of rice – they can withstand crushing forces equivalent to nearly 40,000 times their body weight and survive even being run-over by an automobile.

The reason for the unparalleled ironclad toughness of these beetles resides in the structure of their outer wing covers, each composed of five complicated pieces that are interconnected like the interlocking pieces of a fancy jigsaw puzzle. Given this structure and that of the supports between their wing covers and body, the vital organs in the mid-body of *Phloeodes diabolicus* are well protected from being crushed. Now that the underlying causes of their remarkable resilience are understood, biologists and engineers are teaming-up to apply the beetle's design to produce similarly strong protective linkages for use in bicycles, cars, and even airplanes. Such an approach is not at all uncommon – after all, the beetle's structure is a product of millions of years of biological trial, error and ultimate "best solution," the winnowing-out process of biological evolution that short-cuts any need to re-do what Nature has already accomplished.

In view of the enormous success of insects in the living world, it is perhaps not surprisingly that they were of great interest to Darwin. Indeed, as is shown by his early letters and autobiography, Darwin's young life was often consumed by collecting insects, especially beetles **(Fig. 6-17A to D)**, an avocation that continued throughout his life. Even as a college student at Cambridge he would head out into the countryside whenever he had a chance to search for rare species. Regarding one such sojourn recounted in his autobiography, he penned *"I will give proof of my zeal. One day, on tearing off some old bark, I saw two rare beetles, and* *seized one in each hand; then I saw a third and new kind, which I could not bear to lose, so that I popped the one which I held in my right hand into my mouth. Alas! It ejected some intensely acrid fluid, which burnt my tongue so that I was forced to spit the beetle out, which was lost, as was a third one.*

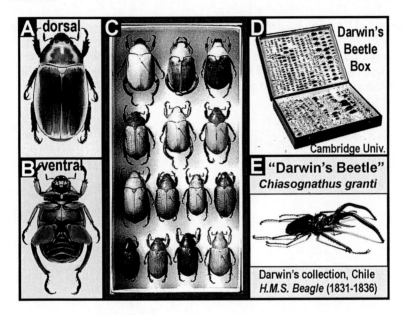

Fig. 6-17 (A-C) Modern beetles, **(D)** Darwin's beetle box, and **(E)** "Darwin's Beetle" *Chiasognathus granti*, specimens of which Darwin collected in Chile during the 1831-1836 voyage of the *H.M.S. Beagle*.

Beetles numbered prominently among Darwin's voluminous collections made during the 1931-1836 voyage of the *H.M.S. Beagle,* one of which (*Chiasognathus granti*) collected in Chile being now known popularly as "Darwin's Beetle" **(Fig. 6-17E)**. Later, in 1858, he wrote his botanist friend Joseph Dalton Hooker *"I feel like an old war-horse at the sound of a trumpet when I read about the capture of rare beetles – is this not the magnanimous simile for a decayed entomologist? It really almost makes me long to begin collecting again."* Even Darwin's children were impressed by his zest for beetles and beetle collecting. As recorded in Darwin's diaries, one of his sons, either George of Francis, asked a grade school pal *"Is your dad's beetle collection as large as my dad's?"* Although there is no real way to know, this rather surprising question may have been uttered because Darwin's son assumed that all fathers were beetle collectors like his – or may have been intended as a bit of "my dad's better than yours" one-upmanship – but in either case it well illustrates the "ignorance of youth" and how youngsters have changed hardly at all over the centuries.

These brief Darwinian tales bring to closure this chapter of the book. Having now reviewed the phylogenetic evolutionary history of protostome coelomates – from lophophorate brachiopods and bryozoans to mollusks, annelids, onychophorans, arthropod crustaceans and insects – we can now turn to the final chapter in this unfolding saga of animal life, the evolution of deuterostome coelomates, the subject of the next chapter of this book and the lineage that led to humans.

CHAPTER 7

BACKBONED ANIMALS TAKE OVER THE WORLD

Deuterostome coelomates – the lineage leading to humans

The term **deuterostome** is a combination of the suffix -stome (from Greek *-stoma*, "mouth") and the prefix *deutero-*, "second" (from Greek *deuteros*, "second," as in the Book of Deuteronomy, literally the "second law," as recounted in the Jewish Torah and the Christian Bible, holy books recounting Moses' second reading of God's Covenants to the Israelites on the plains of Moab shortly before they entered the Promised Land). Thus, the term deuterostome, meaning "second mouth," refers to any of a major lineage of animals **(Fig. 7-1)** in which the second opening in the embryonic blastula becomes the mouth, the first opening becoming the anus. You can easily visualize this by thinking of it as being like the blastula of a protostome animal that during development has flipped on its axis such that the blastopore in-folds ("invaginates") from the bottom rather than the top, the first hole thus becoming the anus and the second hole, where the elongating centrally located primordial gut later penetrates the outer blastula wall, becomes the mouth **(Fig. 7-2)**. Deuterostomes are divided into three groups, the echinoderms, the non-backboned chordates, and the later-evolved highly diverse vertebrate chordates. All of the groups are ancient, all dating from the early Paleozoic, each here considered in turn.

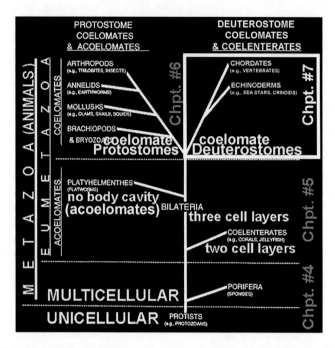

Fig. 7-1 A simplified phylogenetic "Tree of Life" of the Animal Kingdom showing in the white rectangle the part of the tree discussed in **Chapter 7**.

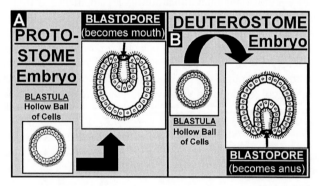

Fig. 7-2 In protostome animals **(A)** the embryonic blastula in-folds at the blastopore to form the tubular gut of the adult animal, the initially formed hole becoming the mouth and the exit pore becoming the anus (for more details, see **Fig. 6.1**). In contrast, in deuterostome animals **(B)** the embryonic blastopore becomes the anus and the exit pore becomes the mouth.

Fossil echinoderms – sea stars, brittle stars, sea urchins and sand dollars and crinoids

Of the three major types of deuterostomes, the **Echinodermata**, "echinoderms" (from Greek *echīnos,* "spiny," plus *derma,* "skin," meaning "spiny skinned") are represented by more than 6,500 living species and about twice as many fossil taxa, making them the second largest grouping of deuterostomes after the backboned chordates. Marine animals found at every ocean depth, from the shallow intertidal zone to the deep-sea abyss, echinoderms are separated into six major taxonomic classes of which the skeletal structures of four, sea stars (the Asteroidea), brittle stars (the Ophiuroidea), sea urchins and sand dollars (collectively, the Echinoidea), and sea lilies (the Crinoidea), have mineralized exoskeletons and thus have good fossil records. In contrast, the body walls of the other two prominent echinoderm classes, sea cucumbers (the Holothuroidea) and sea daisies (the Concentricycloidea) are soft and leathery, their members lacking readily fossilizable hard parts.

Echinoderms are easily distinguished from other deuterostome coelomates by three main characteristics. **(1)** Their adults **exhibit radial symmetry**, a five-part wheel-like arrangement, their bodies not exhibiting obvious bilateral symmetry like other bilaterian coelomates except in their larval stage. **(2)** Their adult bodies are supported by **skeletal structures** made of calcite ($CaCO_3$), a material not used by vertebrates or other chordates where, if present, skeletons are composed of the mineral hydroxyapatite [$Ca_5 (PO_4)_3 (OH)$]. **(3)** Echinoderms possess a biologically unique **"water vascular system,"** a series of tubular canals that pump water in and out of the animal and are connected to their "tube feet" – another echinoderm-only characteristic – that they use for feeding, respiration and to move across surfaces.

Of the four types of echinoderms that have good fossil records, the most familiar to us are probably *sea stars*, represented in the modern biota by some 2,000 described species. Although to some the term "sea star" may be new, members of the group having been long referred to as "starfish," this older moniker is no longer in use given that they are not fish. All such echinoderms are formally assigned to the taxonomical class Asteroidea (from Greek *aster*, meaning "star," plus the Latin suffix *-oid*, "resembling" or "like," the term thus meaning "star like"). Their fossil record dating back to the late Ordovician (about 450 Ma ago), sea stars are the type of animal that usually comes to mind when we think about echinoderms. Their radial five-fold (pentameral) symmetry is readily apparent and because they are relatively common in near-shore marine environments, they are familiar. It is useful to note, however, that sea stars (and brittle stars as well) can have far more than just five arms – some have 10, 15, 20 or even 50 – but they are usually in multiples of 5s **(Fig. 7-3)**.

Fig. 7-3 Three modern sea stars, **(A)** the Atlantic Ocean Royal Sea Star *Astropecten*; **(B)** the Pacific Ocean *Hippasteria*; and **(C)** a 10-armed sea star beneath Artic ice for comparison with **(D)** a modern many-armed ophiuroid brittle star, the Basket Star *Gorgonacephaus* found in Northern Hemisphere circumpolar seas.

Sea stars are predatory, a prime example being their use of the tube feet on the undersides of their arms to pry open bivalve mollusks **(Fig. 7-4)**. Although each of their tube feet exerts only a small pull on a captured clamshell – acting as a small suction cup, a tiny "plumber's helper," to pull the two-hinged shells apart to expose the clam's soft meaty flesh – given enough time and applied en masse they can overwhelm the two strong adductor muscles that keep a clamshell closed. Once the body of their prey is exposed, they then ingest the clam by extruding their stomach out of the mouth situated in the center of the underside of their body, after which the stomach and its meaty fodder retract back inside the sea star. Interestingly, other echinoderms also demonstrate such stomach-extruding evisceration, the giant California sea cucumber (*Parastichopus californicus*), for example, will eject parts of its gut in order to scare and defend against potential predators such as crabs and fish.

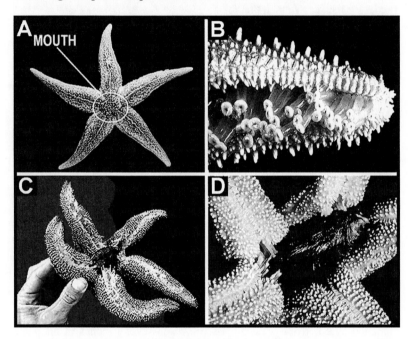

Fig. 7-4 The common sea star *Asterias* showing its mouth centrally located on its underside, its ventral surface **(A)**; its many suction-cup-terminating tube feet extending from the undersides of its arms **(B)**; and its use of such tube feet to pull apart the paired shells of a captured *Mytilus* bivalve clam **(C, D)**.

Although ***brittle stars*** (Ophiuroidea, from Greek *ophis*, "serpent," plus *oura* "tail" and Latin *-oid*, "like" – "serpent tail-like," in reference to their thin spidery arms), also known as "basket stars" **(Fig. 7-3D)**, are represented in the living world by some 2,100 species, they are not nearly as familiar as sea stars, primarily because more than half of their species occur in deep offshore waters. Dating back to the early Ordovician, about 485 Ma ago, living brittle stars mostly occur at depths ranging from 1,500 feet (460 m) to greater than 4 miles (6.5 km). Almost all brittle stars are small, less than an inch (2.5 cm) in diameter, and have five long slender whip-like arms which on the largest specimens can reach up to 2 feet (60 cm) in length, narrow flexible appendages much thinner than the arms of sea stars that they use to crawl slowly across the sea floor. Indeed, they are perhaps best known as "seafloor ecosystem engineers," reshaping the abyssal sediment surface and thereby influencing the distribution of other seafloor life (while at the same time providing welcome fodder to predatory fish, sea stars and crabs).

Some 950 named species of sea urchins and sand dollars – echinoids (from Greek *echinos* meaning "spiny") – are known in the modern world and, like sea stars, they date back to the late Ordovician (about 450 Ma ago). All sea urchins and sand dollars are oval in shape, the *sea urchins* more globose and the sand dollars more flattened, and, again like sea stars, both have their mouths situated at the center of their bottom (ventral) sides. The moveable spines of the urchins, situated on their upper (dorsal) side **(Fig. 7-5)**, are larger and much more evident than those of sand dollars, used by the urchins to pick up food particles that settle onto their upper surface and then step-by-step, spine by spine, transferring the particles to their centrally situated underside mouth. Such urchins typically range in diameter from 1 to 4 inches (2.5 to 10 cm), although the largest species can reach up to 14 inches (36 cm) across. *Sand dollars* are closely allied to the urchins, regarded by some as simply flattened burrowing sea urchins, the Caribbean sand dollar, for example, referred to as the "cake urchin." However, as you no doubt will have noted, the name of their sub-tribe, "sand dollars," seems odd. The "sand" part makes sense, but they look nothing like a dollar bill! Nevertheless, this long-used moniker fits hand-in-glove if you compare them to a silver dollar, a form of American currency dating from 1794 that is rarely used today. The relatively flattened oval body of sand dollars can reach several inches in diameter, their upper surface covered by long fragile fluff-like spines and their five-fold symmetry especially well evidenced by the radiating petal-like imprints of their tube feet on the shells of their fossil representatives.

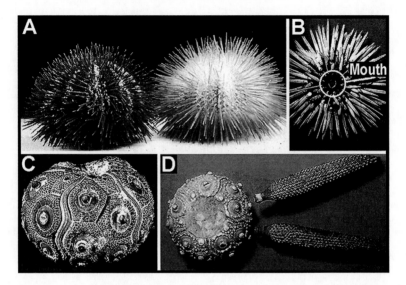

Fig. 7-5 (A, B) The modern sea urchin *Lytechinus*, showing the central location of its mouth on its ventral surface and two Cretaceous (100 Ma) fossil sea urchins, *Leiocidaris* **(C)** and *Cidaris* **(D)**.

Crinoids, the fourth taxonomic class of echinoderms having a good fossil record – their formal name Crinoidea (from Greek *krínon*, "lily," plus the suffix *-oid*, "like," the two together meaning "lily like") coming from the fact that they look rather like flower-topped plants – are best known from the living stalked forms attached to the sea bottom that are commonly called "sea lilies." The fossil record of the group dates from the beginning of the Ordovician (485 Ma ago), the some 600 species living today being generally found at water depths greater than 300 feet (90 m) with some occurring at depths as great as 5.5 miles (9 km). Although the majority of living crinoids are free-swimming, their stem having been lost over evolutionary history, the stalks retained in their deep-sea representatives can be up to 3 feet (1 m) long with some fossil species known to have had slender telephone pole-like stems more than 70 feet (20 m) in length. The stalked forms use their feather-like arms to filter food particles out of the water, their five-fold symmetry being particularly evident in their central "calyx" where the arms are attached to the stem, and by orienting their umbrella-like arms to face an oncoming current **(Fig. 7-6)** they capture food particles from the flowing waters which they then pass along to their calyx-situated mouth. Most live anchored to rocks, but a few can move very slowly across the underlying seafloor.

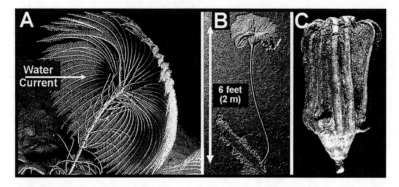

Fig. 7-6 (A) A seafloor-attached modern stalked crinoid that has oriented its umbrella-like feeding arms to filter food particles from the oncoming current and **(B, C)** fossil specimens of stalked crinoids from the Jurassic (150 Ma) Posidonia Shale of southern Germany, the calyx of one **(C)** well-evidencing the characteristic fivefold radial symmetry of members of the group.

When and how did chordates and vertebrates arise?

Echinoderms and chordates – including vertebrate chordates such as you and me and our pet dogs, cats, birds and fish – are deuterostomes and, thus, are undoubtedly interrelated. But how could we be close-cousins of sea stars, not to mention brittle stars, sea urchins, sand dollars and sea lilies? Let's figure it out.

To begin this discussion, we first need to define the term "chordate" (from Latin *chordatus*, "having a [spinal] cord," derived from *chorda* "cord," "string"), a phylum of animals (the **Chordata**) within the larger deuterostome superphylum, the Deuterostomata, which also includes the echinoderms. All chordates, whether or not they have a backbone, share the following four characteristics. **(1)** A **notochord** (from Greek *noton*, "back" plus Latin *chorda* "cord"), a flexible supportive rod made of material similar to cartilage that runs down the back of chordate embryos **(Fig. 7-7)** and that during vertebrate development is replaced by a bony mineralized vertebral column. **(2)** A hollow **dorsal nerve chord** that similarly runs the length of the chordate body, in vertebrates surrounded by the backbone and making up the central nervous system. **(3)** A **post-anal tail**, an appendage that extends beyond the anus at some point in a chordate's life history, a structure that in humans is represented by the short "tailbone" (the coccyx, a remnant of the long-lost tail) that extends from the base of the vertebral column. **(4)** Pharyngeal **gill slits (Figs. 1-6 and 7.7)**, structures present at some stage of a chordate's development, openings in the throat region that in swimming chordates become functioning gills allowing them to extract dissolved oxygen from the surrounding waters but that are also evident in the developing embryos of terrestrial vertebrates **(Fig. 7-7)**. The chordate phylum is divided into three subphyla, the Tunicata, Cephalochordata and Vertebrata.

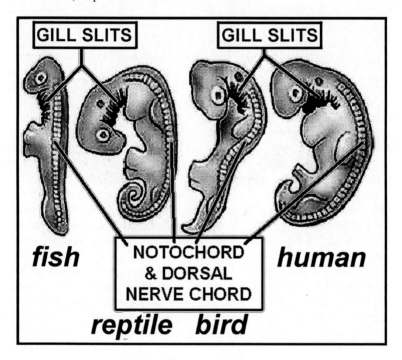

Fig. 7-7 Comparison of the embryos of a fish, reptile, bird and human showing their shared occurrence of gill slits, notochords and dorsal nerve chords.

Sea squirts and free-floating colonial salps are the best-known members of the **Tunicata** (derived from the Latin word *tunica*, the basic Roman attire, used here in reference to the outer "tunic-like" covering of sessile sea squirts). Members of the subphylum are also known as "urochordates" (from Greek *ourá*, "tail," plus Latin *chorda*, "cord"). The body of an adult tunicate is quite simple, being essentially a soft leathery sac that has incurrent and excurrent openings through which seawater is siphoned in and flushed out, the water being filtered for food particles inside the sac-like body. The three taxonomic classes of living tunicates include about 2,100 species.

Members of the **Cephalochordata** (from Geek *kephalo,* "head" plus Latin *chorda*, "cord") – commonly called amphioxus or lancelets – are small marine fishlike animals. The subphylum is small, composed of only 23 living species assigned to only two genera.

In comparison with the tunicates and cephalochordates, the **Vertebrata**, the third subphylum of the phylum Chordata, is enormously numerous and diverse. Indeed, such backboned animals include within their ranks nearly 70,000 described living species of fish, amphibians, reptiles, birds, and mammals. The name Vertebrata (from Latin *vertebra*, "joint") is derived from the Latin *vertere,* "to turn," in reference to the Hox gene-generated multiple vertebrae (the plural of vertebra) that make up vertebrate backbones, structures that are therefore jointed rather than being a single bone or solid spine, a construction that allows backboned animals to move their bodies with bends, twists and turns. In simple parlance, vertebrates are chordates having backbones, whereas tunicates and cephalochordates are chordates that lack backbones. In other words, although all vertebrates are chordates, not all chordates are vertebrates.

So far, so good! We now know the characteristics of chordates and vertebrates, but the two-part question of when and how they arose is still before us. The "when" part, the time of origin of the chordate lineage is firmly settled by the presence of *Pikaia gracilens***,** an extinct early-evolved chordate fossilized in Walcott's Middle Cambrian Burgess Shale of British Columbia **(Fig. 7-8)** with such chordates known also from the even older Cambrian-age Chengjiang Fauna of Yunnan Province, China. *Pikaia*, a flat compressed animal having an expanded tail fin, was a primitive chordate. It lacked a backbone and a structurally differentiated head; averaged about 1.5 inches (3.8 cm) in length; and had a pair of short antenna-like tentacles extending from its frontal region that were augmented by a series of small underlying appendages evidently related to gill slits. Their leaf-like flat bodies were divided into pairs of segmented muscle blocks, evidenced in the fossils by a series of faint parallel lines situated on either side of a flexible, central, rod-like dorsal nerve chord-like structure that runs from the tip of the head to the tip of the tail. *Pikaia* was evidently a slow swimmer, most likely weaving its way through the water by screwing its body into a series of S-shaped, zigzag curves, similar to the swimming movement of backboned eels. Thus, *Pikaia* shows many of the basic characteristics not only of chordates but of vertebrates as well, leaving no doubt that the chordate lineage dates from at least as early as the mid-Cambrian (510 Ma ago).

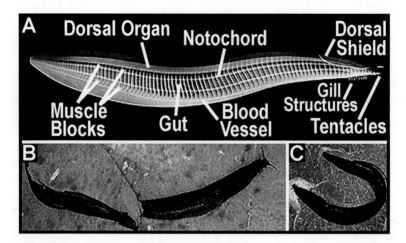

Fig. 7-8 (A) A drawing of the early-evolved chordate *Pikaia gracilens* from the mid-Cambrian (510 Ma) Burgess Shale of British Columbia Canada showing the principle anatomical characteristics evidenced in **(B, C)** Burgess Shale fossil specimens.

Despite this certainty regarding at least the minimal timing of the emergence of chordates, exactly how they arose – their immediate evolutionary ancestry – has, until recently remained something of a mystery. Given that echinoderms and chordates are both deuterostomes, there can be no doubt that the two groups are closely related. Thus, as one might imagine, virtually every conceivable pathway between the two has been proposed. In 1853, German zoologist Johannes Peter Müller (1801-1858) postulated a straight-line descent from **echinoderms** such as sea stars to chordates, based largely on the larval stages of echinoderms. In 1900, University of Utah biologist Ralph Vary Chamberlain (1879-1967), noting the presence of both primitive and advanced characteristics in **cephalochordates**, lancelets, suggested that they were the immediate forerunner of chordates. In 1928, British zoologist Walter Garstang (1868-1949), hypothesized that **tunicates**, urochordates, gave rise to chordates, basing his concept chiefly on their tadpole-, amphibian-like larvae, an idea again promoted in 1955 by English marine biologist Norman John ("Jack") Berrill (1903-1996). In 1959, Harvard University vertebrate paleontologist Alfred Sherwood Romer (1894-1973) proposed that ancestral deuterostomes were sedentary tentacle-feeders like modern **hemichordates** such as acorn worms, a group that like vertebrates exhibits pharyngeal gill slits. Most recently, in 1981 and again in 1997, British Museum paleontologist Richard P.S. Jefferies (1932-2020) championed the hypothesis that chordates arose from **calcichordates**, a fossil-based sister group of chordates known earliest from the Middle Cambrian of Utah (an interesting suggested linkage disproven in the early 2000's).

Of these, the "winner" is Romer's Hemichordate Hypothesis, the notion that chordates were derived from hemichordate marine deuterostomes, a concept supported by anatomical, developmental, genetic and fossil evidence. Such hemichordates (from Greek *hemisus*, "half" plus Latin *chordatus*, "having a [spinal] cord," the term meaning "half-chordates") are generally worm-like deposit feeders (detritivores) that like other chordate deuterostomes have pharyngeal gill slits and most have a dorsal nerve chord although they lack a notochord. Thus, hemichordates are now accepted to be the closest existing relatives of vertebrate chordates with their existence by the mid-Cambrian – required for them to be a plausible precursor of the vertebrate lineage and known since Walcott's 1911 discovery of acorn worms in the Burgess Fauna – having been reconfirmed and amplified quite recently by yet another discovery from the Burgess Shale.

This new find of an ancient hemichordate, *Gyaltsenglossus senis* – reported from the Burgess Shale in 2020 by Royal Ontario Museum invertebrate paleontologist Jean-Bernard Caron and Université de Montréal integrative biologist Christopher B. Cameron – is particularly well preserved. Despite being less than an inch (about 2.5 cm) in length, it shows such fine anatomical structures as a crown of six feeding arms covered by short tentacles. Moreover and most importantly, its anatomy combines characteristics found in the two living groups of hemichordates, the Enteropneusta (acorn worms) and the Pterobranchia, small worm-shaped hemichordates that live in self-secreted tubes on the ocean floor that are well known from mid-Cambrian- to Carboniferous-age fossils known as graptoloids.

Prior to this recent report, the more than 170-year-long confusion regarding the evolutionary roots of chordates was markedly hampered by the understandably incomplete fossil record of hemichordates, a result of them being almost entirely soft-bodied and thus unlikely to be preserved in most fossil-bearing deposits. Indeed, only a handful of fossil hemichordates are known that preserve the soft tissue anatomy crucial to understanding their early evolution. Fortunately for science, *Gyaltsenglossus* fills the bill. Thus, the Caron-Cameron discovery and study of *Gyaltsenglossus* represents a major and perhaps final turning point in the debate, in part because it shows that early-evolved hemichordates possessed features of both enteropneusts and pterobranchs, a suite of characteristics that may have given them a flexibility of life-styles that promoted their success. In short, this new find clarifies the hemichordate origin of vertebrates, a discovery that may have once and for all put behind us the long-running debate as to how vertebrates came to arise.

To close this part of the discussion and illustrate the human-side of science, I here recount a brief vignette about how I first met Professor A. S. Romer whose Hemichordate Hypothesis has been borne out by *Gyaltsenglossus* and other telling lines of evidence. In the spring of 1962 when I was a first-year graduate student, my professor, Elso S. Barghoorn, took me across the street from our lab-building to show me Harvard's Museum of Comparative Zoology. As we approached the MCZ, he led me over to meet an elderly gentleman sitting on the stairs near the street smoking a cigarette. Barghoorn introduced us and, without thinking, I loudly blurted out *"Al Romer!?"* I was stunned, astounded – I had read Romer's textbook on Vertebrate Paleontology when I was a college student and here he was in the flesh. Wow! Later, when we became friends, he enjoyed kidding me about this little episode, always smiling, always kind. Though I remembered my enthusiastic outburst with embarrassment, he remembered it fondly, telling me how pleased and honored he was that a fledgling grad student had ever heard of him! Professor Romer was a splendid scientist and a warm generous human being.

Fish – the earliest-evolved chordate vertebrates

Backboned animals, the Vertebrata, make up the largest subphylum within the chordates and are the most morphologically complex. In addition to the typical characteristics of chordates, all vertebrates possess a skull (a "cranium"*)* which encases the brain, and a backbone (a vertebral column) which protects the dorsal nerve chord and internal organs as well as providing structural support to the body. In addition, virtually all vertebrates also have a hinged jaw which allows them to capture food in a highly effective way; amniote eggs, even in mammals, which includes a protective inner membrane through which gases and nutrients can be transferred to the embryo during its development; limbs, present either as fins, in swimming vertebrates, or as evolved legs that enable effective movement on land; and oxygen-breathing lungs, derived from pharyngeal gills.

Vertebrates are divided into eight taxonomic classes, four of which are fish or fish-like animals – three extant and one extinct – with the other four represented by living terrestrial four-limbed tetrapods. Listed in their evolutionary sequence – the order in which they are considered here

– the eight classes include **(1) jawless fish** (the Agnatha), **(2) cartilaginous fish** (the Condrichthyes), **(3)** extinct **armored fish** (the Placodermi), **(4) bony fish** (the Osteichthyes), **(5) amphibians** (the Amphibia), **(6) reptiles** (the Reptilia), **(7) birds** (the Aves), and **(8) mammals** (the Mammalia).

The **Agnatha** (from Latin a-, "without" plus Greek *gnathos* "jaw," meaning "without jaws") are represented in the modern world by some 650 species of lampreys and hagfish, jawless fish that are included in the superclass Cyclostomata. Although agnathans have a skull and the basic components of vertebrae, they lack a well-defined mineralized backbone, a "vertebral column." Both lampreys (blood-sucking parasites) and hagfish (vacuum cleaner-like scavengers) also lack jaws **(Fig. 7-9)**, their mouths having horny epidermal structures that function as teeth. The earliest known fossil agnathans are lampreys that date from Walcott's mid-Cambrian (510 Ma) Explosion of Life.

Fig. 7-9 Living agnathan jawless fish, **(A)** blood-sucking parasitic lampreys and **(B)** an ocean-floor scavenging hagfish.

The **Condrichthyes** (from Greek *khondros*, "cartilage" plus *ikhthus*, "fish," hence "cartilaginous fish") – today's sharks, skates, rays and sawfish – includes about 1,000 living species. All have gills and are characterized by an internal skeleton (endoskeleton) made of cartilage. Such cartilaginous fish are jawed vertebrates having paired fins, paired nasal openings (nares), scaly skin, and a heart with its blood-pumping chambers arranged in series. Dating back to the mid-Ordovician (465 Ma), ancient condrichthyans are probably best known from their fossilized teeth. Sharks, for example, most of which feed on fish, have multiple rows of many-cusped teeth arranged such that if those in the front row become detached, lodged in their prey, they are soon replaced by those in the row immediately behind. Their teeth are made of calcium phosphate [Ca_3 $(PO_4)_2$], calcified dentin, a very tough material. Moreover, because a shark can shed its teeth many hundreds of times throughout its life – and because fossil shark teeth can be large and stunningly impressive – they are abundant, identifiable and readily collectable.

The **Placodermi** (from Greek *pláka*, "plate" plus *derma,* "skin," meaning "platy skinned"), the first fish to develop the pelvic fins precursors of the hind-limbs of vertebrate tetrapods, is a class of extinct armored fish. Formerly also referred to as ostracoderms, placoderms are perhaps best known from the Devonian platy skinned fish *Dinichthys*), fossils of the class ranging from the early Silurian (440 Ma) to the end of the Devonian (360 Ma). Like their chondrichthyan cartilaginous cousins, placoderms were among the first jawed fish, their jaws likely evolved from their front-most gill arches. Articulated bony plates covered the head and thorax of placoderms, the rest of the body being either scaled or naked depending on the species. With but one exception (the late Silurian 425-Ma-old genus *Entelognathus* from China which exhibits a bony jaw like that of modern bony fish) the jaws of placoderms were simple, consisting of but a single bone.

The **Osteichthyes** (from Greek *osté[on]*, "bone" plus *ikhthus*, "fish" – "bony fish") have gills, an endoskeleton made of bone, and a "swim bladder," an internal organ filled with air (or sometimes an oily fluid) that allows them to float and helps them control their depth in the water column. Of the nearly 70,000 living species of living vertebrates, nearly half are osteichthyans of the superclass Pisces (jawed fish), ray-finned members of which are known as actinopterygians (from Latin *actino-*, "having rays," plus Greek *ptérux*, "wing, fins," so named because their fins are webs of skin). The Osteichthyes, the largest taxonomic class of vertebrates, includes more than 95% of all fish species living today. Indeed, included here are virtually all of the fish familiar to us, today's trout, perch, bass, salmons, tuna, herrings, pufferfish and even flounders. The earliest known fossil bony fish (*Guiyu oneiros*, dating from the late Silurian, about 430 Ma ago) exhibits a combination of both ray-finned and lobe-finned features, the latter set of traits exhibited in the living world by the lobe-finned coelacanth *Latimeria*.

The modern **coelacanth** genus *Latimeria* **(Fig. 7-10A, B)** includes both of the two-surviving species of this lineage that bridged the evolutionary gap between aquatic vertebrates and their land-dwelling descendants, a development dating from the mid-Devonian (about 400 Ma ago). During the Devonian, however, there were myriad species of such lobe-finned fish, all formally referred to as crossopterygians (members of the subclass Crossopterygii from Greek *krossoí*, "tassels, fringe" plus *pterýgion*, "little wing or fin," hence "tasseled fin"), perhaps the best-known example being *Osteolepis panderi* **(Fig. 7-10C, D)** that lived in shallow nearshore brackish and freshwater lakes and ponds. From time to time, typically seasonally, the local setting would dry out, forcing *Osteolepis* to use its muscular lobed fins to slither across the mud from one drying pool to the next. If that pond then also dried to a crisp it would burrow headfirst into the underlying mud, coat its encasing cavity with mucous, and hibernate ("aestivate") in its cocoon-like bulbous chamber until the next wet-season restored the pond. But if the pond were never restored or was sealed by an overlying layer of volcanic lava or in-falling ash, the hibernating dormant fish would be fossilized. Thus, it not surprising that we have a good record of the anatomy and life-style of such lobe-finned fish – a record so complete that we can compare the homologous muscles and bone structure of their lobe fins to those of early land-invading amphibian tetrapods **(Fig. 7-11)**.

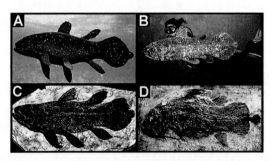

Fig. 7-10 (A) The West Indian Ocean coelacanth *Latimeria chalumnae* from coastal waters off the Comoros Islands northwest of Madagascar and **(B)** the other living lobe-finned species, the Indonesian coelacanth *Latimeria menadoensis*, compared with **(C, D)** the Devonian (400 Ma) fossil lobe-finned fish *Osteolepis*.

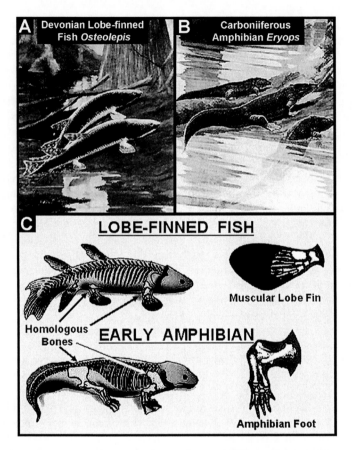

Fig. 7-11 (A) The Devonian lobe-finned fish *Osteolepis* compared with **(B)** the early-evolved Carboniferous to Permian amphibian *Eryops*, and **(C)** a comparison the limb-associated homologous bones of lobe-finned fish with those of early land-invading amphibian tetrapods.

The credit for establishing the linkage between coelacanth crossopterygian fish and early amphibians belongs to fossil fish expert Colin Patterson (1933–1998) of London's Natural History Museum with whom I became a close friend in 1989 when I spent a year at the museum as a visiting scientist. Our friendship stemmed mostly from the fact that we two were the only scientists in our part of the museum who worked on weekends (and also, perhaps, because it was I, a mere "visiting Yank," who figured out how to defeat the electrical outages that plagued our part of the building).

As Patterson recounted to me, here is how he made major inroads into solving the conundrum of the origin of land animals. Years before, he had heard about a particularly large fish (that turned out to be the coelacanth *Latimeria*) being occasionally snared in the nets of trawlers off the Comoros Islands northwest of Madagascar. Intrigued, he journeyed down to have a look. He there discovered that once caught, the fishers simply threw the fish back into the sea because they were "not good eating." Patterson obtained a couple of specimens – for free, their catchers being glad to get rid of them. He then brought them back to the museum where he studied their muscles and bone structure, and, because as a serous scholar he was well aware of what was then known about Devonian fish, he figured out how they could bridge the water-to-land vertebrate gap. Next, Patterson put in place a fine museum exhibit of the coelacanth-to-amphibian transition. Other museums soon heard of this breakthrough and the price of these previously unwanted large fish skyrocketed, from $50, to $250, to more than $5,000 each, the fishers enjoying a bonanza. Ultimately, however, the market became glutted – every museum that wanted a coelacanth had one. The market price of coelacanths then plummeted to near nothingness and the trawlers once again started discarding the coelacanths back into the sea.

Patterson's little story carries a couple of useful take-home lessons. First, as the French microbiologist and chemist Louis Pasteur (1822-1895) taught us in 1854, *"fortune favors the prepared mind,"* a mind such as that of Colin Paterson who was fully knowledgeable about the relevant fossil record when he set out to tackle the fish to amphibian evolutionary transition. Second, it is naive to imagine that science and the culture in which it operates are totally disconnected. Yes, of course, new finds make their mark, but science and the society in which it is carried out are inextricably linked, a linkage that is sometimes difficult to discern and other times (such as the rise and fall of the going-price of coelacanths) abundantly obvious.

To these, a third object lesson – a caveat – should be added, for despite how sensible and fact-supported Patterson's *Latimeria*-based synthesis appears to be, it is not the whole story. As the old adage has it, *"Science never sleeps."* Unanticipated new finds, new evidence, new facts can augment a widely accepted interpretation at virtually any time. In this instance, a case in point was the 2004 discovery of a *Latimeria*-like but even more primitive fossil crossopterygian, the Late Devonian (375 Ma) lobe-finned fish *Tiktaalik roseae* discovered in the wilds of Ellesmere Island in Nunavut, Arctic Canada. The find was made by paleontologist, evolutionary biologist and popular science writer Neil Shubin (University of Chicago and Chicago's Field Museum of Natural History) and two colleagues, paleontologist Farish Alston Jenkins (1940-2012, Harvard University) and vertebrate paleontologist Edward B. "Ted" Daeschler (Drexel University and the Academy of Natural Sciences, Philadelphia). Although, like *Latimeria*, *Tiktaalik* is certainly a lobe-finned fish, complete with scales and gills, its flattened head looks more like that of a crocodile, a reptile, not a fish! And though its fins like those of most fish have thin supporting ray bones, they also have more robust interior bones that would have enabled *Tiktaalik* to prop itself up in shallow water and use its limbs for support as do land-inhabiting tetrapods (four-legged animals). These distinctive fins and a suite of other characteristics akin to those of tetrapods set *Tiktaalik* apart, evidencing the mix of features to be expected between swimming fish and their land-inhabiting descendants. Thus, *Tiktaalik* and its close relatives could well be the common ancestors of the vertebrate terrestrial fauna, the amphibians, reptiles, birds, and mammals of the world today.

Regardless of whether Patterson's *Latimeria* or Shubin's *Tiktaalik* explanation of the origin of land animals ultimately holds sway, there is no doubt that the earliest land animals evolved from lobe-finned fish. Nevertheless, before we move on to consider animal life on land – tetrapod amphibians and their chordate vertebrae close cousins – it would be remiss not to recount a few additional miscellaneous facts about living fish and their ancient aquatic ancestors. Here is a list of ten that may tickle your fancy:

(1) Catfish have over 27,000 taste buds whereas humans have only 9,000.

(2) Fish typically have quite small brains relative to body size compared with other vertebrates, typically far less than 1% the brain mass of a similarly sized bird or mammal.

(3) Fish have a specialized sense organ called a lateral line which works much like radar and helps them wend their way through dark or murky waters.

(4) It was not until 1853, when aeration and filtration of water was first worked out in England that aquaria could be built and people were able to keep fish as indoor pets.

(5) Sturgeons have inhabited the northern U.S. Great Lakes for 10,000 years, since the end of the Neogene (Pleistocene) "Ice Age"; they never stop growing, can live as long as humans and can reach an adult human-sized length of 6 feet (2 m).

(6) Every fall, marine Chinook, Coho, Pink and Atlantic salmon migrate upstream to spawn, traveling to lay their eggs through freshwater rivers to the same location where they were born.

(7) When searching for a mate, male freshwater Drum Fish (also called "croakers," "grunters" and "grinders") make a grumbling, rasping sound by vibrating their muscles across their swim bladders, a great attractor of female Drum Fish.

(8) Long-nose Gar extant for about 100 million years, have thick, interlocking, armor-like scales that protect them from predators, razor sharp teeth, and mouths that are twice the length of their heads.

(9) American Eels, covered with a layer of mucus that makes them "as slippery as an eel," are nevertheless able to absorb oxygen through their skin and gills so that they can move over mud, rock, muck and wet grass as they slither from one locale to another.

(10) The largest known living fish, the great white shark *Carcharodon*, grows to a length of about 20 feet (6 m), weighs about 4,000 pounds (1,800 kg) and has teeth about 2 inches (5 cm) long. In contrast, the largest fossil shark – the extinct Neogene (Eocene to Pliocene, 23 to 3 Ma) *Otodus megalodon* (meaning "big tooth") and more commonly known simply as "Megalodon" **(Fig. 7-12)** -- was far larger. Megalodon adults ranged in length from some 40 to 60 feet (12 to 18 m), weighed 66,000 to 140,000 pounds (30,000 to 63,500 kg) and had teeth as long as 7 inches (18 cm).

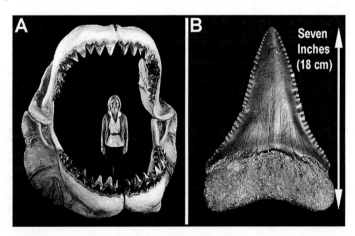

Fig. 7-12 (A) The largest fossil shark, the extinct Neogene (Miocene to Pliocene, 23 to 3 Ma) "Megalodon" having a gape larger than 6 feet (2 m) and **(B)** 7 inch- (18 cm-) long teeth.

Amphibians and reptiles –
animals dominate the landscape

The remaining four classes of backboned animals are all tetrapods, four-limbed chordate vertebrates. Of these four groups, the earliest to evolve was the **Amphibia** (from Greek *amphibios* meaning "to live a double life"), a name coined in reference to the fact that that members of this class – frogs, toads and salamanders, for example – spend part of their lives in water and part on land. Including about 7,000 living species, of which nearly 90% are frogs, amphibians are initially water inhabiting and, as they mature, become terrestrial. Although in the adult form most of them have lungs, they can also breathe through their thin moist skin. All amphibians are cold-blooded (ectothermic) vertebrate chordates, meaning that like fish and reptiles they take in heat from their environment rather than regulating their body temperature internally. In addition, all are distinguished from other vertebrates by having an aquatic gill-breathing larval stage followed (typically) by a terrestrial lung-breathing adult stage. In frogs, for example, the post-hatching larval stage is called a "tadpole" that has a tail and gills. As it matures, the tadpole develops first two legs, then four; its tail lengthens and it grows into a short-tailed "froglet"; and only after this prelude does it grow into a full-fledged frog. The earliest well-known fossil amphibian *Ichthyostega* occurs in late Devonian (363-Ma-old) deposits of Greenland whereas the oldest amphibian discovered to date, *Elginerpeton*, occurs in slightly older Devonian rocks of Scotland.

Amphibians **(Fig. 7-13A, B)** were the first tetrapods to venture onto the land and, thus, were the initial "movers and shakers" that led to the eventual establishment of animal-life on land. By making this major break-through they accomplished a lot, it most importantly providing them access to the previously untapped food resources available on the land surface. Nevertheless, they failed to complete the task – their adults could walk or even hop and jump across the landscape, but their juvenile forms were tied to living in water. In other words, they accomplished only half the task – and, we all know, "a job half-done is a job undone!" Why did they fail to complete the job?

The answer is simple and here again, as it was for the earliest land-inhabiting rhyniophyte spore-plants, is tied to reproduction. Most amphibians reproduce in freshwater, their soft gelatinous masses of shell-less eggs **(Fig. 7-13C, D)** deposited on the shallow bottom of a pond with a male then swimming over the egg mass and spraying out sperm. Thus, unlike all later-evolved tetrapods, fertilization in amphibians occurs outside the body and their soft eggs also differ markedly from those of

their evolutionary descendants that dominate the landscape today, the
reptiles and birds with their hard-shelled eggs and almost all mammals
with their internally protected developing embryos.

Fig. 7-13 Amphibians, **(A)** a bright-eyed frog and **(B)** the barred tiger salamander,
the official state salamander of Kansas. Amphibians are tied to water to reproduce,
their masses of soft gelatinous eggs, such as those of frogs **(C, D)** being too fragile
to survive on dry land.

Unlike the eggs of later-evolved reptiles, birds and mammals,
eggs that are termed "amniotic" with such tetrapods thus being known
collectively as amniotes (from Greek *amnion*, "membrane surrounding the
fetus"), amphibian eggs are "*an*amniotic" (meaning not amniotic). Thus,
amphibian eggs are much like fish eggs, such eggs having a structure
composed of a series of jellylike layers that help to inhibit the desiccation
and resulting death of the developing embryo, layers that in turn are
surrounded by an enclosing membrane through which outside oxygen
passively diffuses and enables the developing frog, toad or salamander to
breathe. Because amphibian eggs lack either a thick outer shell – as in
reptiles, birds and primitive monotreme mammals such as the egg-laying
platypus – or an embryo-encasing body wall, as in all mammals except
monotremes, the developing eggs of amphibians must be surrounded by
water or situated in a very damp location to prevent them from drying out.

In contrast with the eggs of anamniotic amphibians, the amniotic egg of later-evolved tetrapods contains quite a number of distinct features. At the innermost part of such eggs the embryo is surrounded by amniotic fluid, a watery protective liquid – that, interestingly, is simply an evolutionarily advanced version of the watery milieu required by amphibian eggs – the amniotic fluid being contained within a membrane called the amnion. The closely associated allantois membrane enables waste and gas exchange and, together with the nutritious proteinaceous egg yolk, connects to and partially surrounds the embryo. Enclosing all of these is yet another membrane, the chorion that separates them from the albumin, the familiar "egg white" of a chicken egg. In reptiles, birds and monotreme mammals, all of these components are enclosed within an outer hard shell, such shells being calcified in most lizards, snakes, turtles, crocodiles and birds. In more-evolved marsupial and placental mammals, however, animals that give birth to live young, the developing fetus is cushioned in a non-calcified flexible layer, the amniotic sac that ruptures during late-stage pregnancy. In humans, this event typically occurs at the beginning of or during labor, the explanation for the so-called "water breaking" that occurs soon before the newborn enters the world.

Why did the development of the amniotic egg matter? This, too, is easy to understand. Because of their more primitive mode of reproduction, frogs, toads, salamanders and their amphibian ancestors were bound to water, a requirement cast aside by the development of the amniotic egg that permitted their bearers to venture into habitats far distant from pooled water occupied by thriving communities of seed-plants and their accompanying large populations of insects, a previously untapped food source. Thus, the "eaters," the amniote vertebrates, followed the "eatees," the plants and their associated insects. Such amniotes needed to develop new behavioral patterns, such as "nesting" to protect their eggs and "flocking" to protect themselves from other amniote predators, but such new ways of living were a small price to pay for the newly available rich supply of fodder.

The earliest known amniote tetrapods were members of the **Reptilia** (from Latin *reptilis*, "creeping or crawling"), a group that includes lizards, turtles, snakes, crocodiles and their kin, all of which have air-breathing lungs and all, like amphibians, being cold-blooded ectotherms. Numbering about 11,500 living species, reptiles are the largest vertebrate group after fish and they range widely in size, from gecko lizards that are under an inch (2.5 cm) in length, to saltwater crocodiles up to 20 feet (more than 6 m) long. All are basically terrestrial, even such seemingly aquatic reptiles as crocodiles and marine turtles that return to sandy or muddy beaches to lay their eggs. All have a scaly watertight skin devoid of either sweat glands or other moistening glands and thus being dry and rough, a snake's scales being composed of keratin, the material that makes up human fingernails. The bodies of reptiles are divided into a head, neck, trunk, and tail, and although present on every continent except Antarctica, because of their cold-blooded metabolism they occur most commonly in the warmer regions of the globe. The earliest known fossil reptile is *Hylonomus lyelli,* particularly notable as the first animal known to have fully adapted to life on land. *Hylonomus* lived about 315 million years ago, during late Carboniferous.

The Reptilia, a taxonomic class in the system of scientific nomenclature, is only a part of a larger taxonomic category, the **Synapsida** (from Greek *syn-*, "together," and -*apsid*, "arch," referring to the holes, "temporal fenestra," low in the skull roof behind each eye that leave a bony arch below). The Synapsida includes not only all mammals but also numerous superficially "reptile-like" extinct animals that are actually more closely related to mammals than to other non-mammalian amniotes such as dinosaurs and birds. This distinction is important because in years past such supposedly "dinosaur-like" synapsids were described as "mammal-like reptiles," an outdated terminology no longer in use given that synapsids are no longer considered to be reptiles. Instead, early-evolved synapsids are now more correctly referred to as "stem mammals" or "proto-mammals." Such non-mammalian synapsids existed for over 80 million years before the first mammals evolved during which they evolved a great array of shapes, sizes and ways of life including large saber-toothed carnivores, herbivores having turtle-like beaks, carnivores and herbivores having tall sails extending from their backsides, and many others. Synapsids are one of the two major groups of amniotes, the other being the sauropods, the group that includes reptiles and birds.

Let's now return to the amniote non-synapsid (sauropod) Reptilia, rightful owners of numerous "claims to fame," perhaps most notably being among the longest-lived animals on the planet. Though it may be news to you, the pet turtle you harbor at your home can live up to 80 years while larger species in the wild can easily live 100 years or more. Indeed, the world record-holder for longevity among all types of terrestrial animals is the Aldabra Tortoise, native to the Aldabra Atoll in the Seychelle Islands northeast of Madagascar in the Indian Ocean, one specimen of which – well documented because it was housed at Alipore Zoological Gardens in Kolkata, India – hatched at Aldabra in 1750 and died in 2006 at the ripe-old-age of 255! This particular tortoise was also a real heavyweight, tipping the scales at 550 pounds (250 Kg). But the Aldabra species is only the second-largest land tortoise known, that prize being held by the Galapágos tortoise, well known from its 1835 Darwin-recorded presence in the Galapágos Islands some 600 miles west of Ecuador, giant tortoises that can weigh up to 920 pounds (420 Kg). Like their Aldabra cousins, the Galapágos tortoises are also long-lived, averaging over 100 years with the oldest on record living to 152. Such longevity is exhibited by other reptiles as well – alligators, for example, alligators, for example, can live to an age of more than 70 years.

Similarly, if surprisingly, the sex of numerous types of reptiles is determined not primarily by their genes, as it is for humans, but by the temperature of their surroundings before they hatch. As an example, for turtles the ambient temperature outside the egg, during the development of the embryo, will determine whether a hatchling is male or female. The gender-determination of baby alligators is much the same, the temperature in which the eggs develop dictating the newborn's sex with eggs exposed to temperatures at or below 86°F (28°C) becoming female and those experiencing temperatures above 93°F (34°C) becoming male.

Dinosaurs take over the scrum as ectothermy is cast aside and birds arise

Clearly, living reptiles are truly fascinating, but reptiles especially stand out because the huge now-long-extinct roaring fearful beasts that we so admire – the lumbering, prancing, leaping dinosaurs that ruled the Mesozoic – were all reptiles as well. Fossils of the **Dinosauria** (from Greek *deinos*, "fearfully great," plus *sauros*, "lizard" – "fearfully great lizards") are found on all seven continents, the group and its reptilian close-cousins dominating Earth's land, sea and air for some 180 million years, dating from the mid-Triassic (235 Ma) to the end-Cretaceous (65 Ma) Chicxulub Impactor event (**Chapter 3**). Dinosaurs are divided into two great subgroups named for their pelvic and leg-bone-structure, the Saurischia ("lizard-hipped") and the Ornithischia ("bird-hipped") dinosaurians. The saurischians include carnivorous dinosaurs such as *Tyrannosaurus rex* as well as herbivorous plant-eating types such as *Apatosaurus* (known earlier as *Brontosaurus*), whereas the ornithischians include the remainder of the plant-eaters such as *Triceratops*. Rather surprisingly, birds evolved from the "lizard-hipped" rather than the "bird-hipped" dinosaurian lineage.

Considered as a whole, dinosaurs are highly varied, known from over 900 genera and many more individual species. Although all were evidently egg-laying nest-builders and some were bipedal, ambling about on their two hind legs, the great majority were quadrupeds, using all four legs to roam. Many of the best known and best-preserved dinosaurs had bony armor and spines, and though horns and frontal crests were common to all groups of dinosaurs they were particularly evident in herbivorous vegetarians where they served to ward off their omnipresent carnivorous predatory brotheran (and may also have been important in attracting mates).

Although many dinosaurs were quite small – *Xixianykus*, for example, being only about 20 inches (50 cm) in length – the ones known best to us were huge, evidently the largest being *Argentinosaurus*, known from Argentina, South America, which measured over 130 feet (40 m) from nose to tail-tip, as long as four fire engines parked front to back. Moreover, and not surprisingly, dinosaurs are especially well known as being fearsome fighters, especially the carnivores attacking and attempting to devour their co-existing meal tickets, the more placid plant-eating herbivores. Here the best known such predator-prey pair is *Tyrannosaurus rex* and *Triceratops,* Cretaceous dinosaurs that co-existed for a few million years. They, together with their giant dinosaurian relatives have molded our views of these wondrous beasts, but this particular pair is locked in our memory banks for it is these two dueling dinosaurs that are depicted **(Fig. 7-14)** in front of museums, inside of museums, on paperweights, on T-shirts, and even on cereal boxes! Indeed, since their discovery in the early 1800's, fossil dinosaur skeletons have been major attractions at museums worldwide and dinosaurs have become embedded in human culture – a fascination well evidenced by popularity of films such as *Jurassic Park.*

Fig. 7-14 The Cretaceous dueling dinosaurs T*yrannosaurus rex* and *Triceratops* **(A)** in front of and **(B)** inside museums, on **(C)** paperweights and **(D)** T-shirts, **(E)** even on cereal boxes!

Yes indeed, reptilians dominated the Mesozoic landscape, but they also ruled the sea and sky. Swimming reptiles known as plesiosaurs, ichthyosaurs and mosasaurs, vied with sharks to be the dominant carnivores of the oceans, the longest such plesiosaur being *Elasmosaurus* up to 46 feet (14 m) long, half of its length being its neck composed of as many as 75 vertebrae (in comparison to 7 or 8 neck vertebrae in humans). *Elasmosaurus* had a small head, strong jaws and sharp teeth, four long paddle-like flippers, and a pointed tail.

Fig. 7-15 *Plesiosaurus dolichodeirus* on display at London's Museum of Natural History, accompanied by a plaque (lower left) crediting its collector, Mary Anning.

Despite its gigantic size, *Elasmosaurus* is not the best known and most admired plesiosaur on record. Instead, that distinction goes to a splendid specimen of *Plesiosaurus dolichodeirus* on permanent display at London's Museum of Natural History (**Fig. 7-15**) collected in December 1823 by the pioneering English paleontologist Mary Anning (1799-1847) from the Jurassic limestones and shales that form the prominent cliffs that flank the English Channel along southwestern-most England. Anning (**Fig. 7-16**) was "one-of-a-kind," a fossil collector known throughout Europe for her finds of plesiosaurs, ichthyosaurs, pterosaurs and numerous other Mesozoic fossils but also for her intrepid courage and perseverance. She came from a working-class family and, like others brought up under the early-1800's English "Class System," she had an extremely limited formal education. Coupled with that, she and her family were shunned as religious dissenters – not followers of the Church of England – with her fossil finds of huge animals so vastly different from those living today being at odds with widely accepted Biblical wisdom. Finally and perhaps most importantly, because of her gender her ground-breaking contributions were often overlooked or ignored by the all-powerful "Upper Class" scholarly elite, a situation that might well have been different had she been eligible to be elected to the Geological Society of London – which, as a female, she was not. Clearly, Mary Anning was a heroic contributor to our knowledge of extinct, wondrous, ancient reptilians!

Fig. 7-16 Pioneering English paleontologist Mary Anning.

If you had gazed across the Mesozoic landscape you would have seen ferocious dive-bombing reptilian pterosaurs in the skies above – this being well before they were supplanted by birds – the largest pterosaur being *Quetzalocoatlus* that had a wingspan of up to a whopping 43 feet (13 m) and may well have been the largest flying creature of all time. Interestingly, despite its size it weighed no more than about 220 pounds (100 kg), the relatively small weight of its body mass in comparison with its huge wingspread enabling its aerial deviltry.

While there no doubt that reptiles gave rise to birds – the linkage being so strong that a great many scientists regard birds as simply an advanced lineage of "flying feathered reptilians" – there are other differences between reptiles and birds besides the presence or absence of feathers. The most notable among these is the difference between their genetically determined abilities to regulate internal body heat, their "thermoregulation." Like their fish and amphibian ancestors, reptiles are cold-blooded ectotherms (from Greek *ectos,* "external" plus *thermē* "heat" – "outer heat") that absorb their body heat from the surrounding environment. In contrast, birds and mammals, the two more advanced chordate lineages, are warm-blooded endotherms (from Greek *endo,* "within" plus *thermē* "heat" – "internal heat"), animals such as ourselves in which body heat is determined by internal metabolic processes. Why did warm-bloodedness, endothermy, win the evolutionary competition?

In comparison with ectothermy, many aspects of endothermy are advantageous, most importantly the resulting decrease in an animal's vulnerability to fluctuations in the outside temperature. The overall rate of an animal's metabolism doubles for every rise of 18°F (10°C) in temperature. Thus, the internally controlled relatively high body temperature of warm-blooded endotherms results in faster metabolism and greater stamina than that of ectotherms, allowing endotherms to quickly and continually replenish their muscles with chemical energy (mostly in the form of adenosine triphosphate, ATP) and rapidly break down the waste-products their muscles produce. Moreover, to optimize muscular movement, whether for feeding or self-protection, most animals need to maintain their core body temperature within a relatively narrow range.

The following story illustrates the point. During the summer after my third year in college, I, like many geology majors, enrolled in a "summer geology field course," in my case stationed in Sheridan, Wyoming. One rather chilly morning I was assigned the job of leading a small student-field-party sent out to map a particular set of geological strata. Seven had been assigned to the group, but one, Mohamed T. El-Ashry, an Egyptian graduate student studying at the University of Illinois, decided that it was too early in the morning for him to do geology, so he stayed behind in the field vehicle to catch-up on his sleep – a decision that did not sit well with the rest of the group.

The other six of us headed off and by about 8 am I was leading the group over a small slope (in the Triassic Spearfish Formation) and was about to jump down to the ledge below when I spotted a coiled six-foot rattlesnake directly in my path. I then led the group around the slope and we approached the snake from the side. My father was brought up on a ranch in Wyoming and had told me when I was a youngster that if I ever came across a rattlesnake I had to kill it because otherwise it would poison one of the calves in a rancher's herd. The six of us then pinned the snake down with a fusillade of rocks and I used my geology hammer to chop off its fearsome fang-filled head and bury it deep in the soil (as my father had instructed me, the muscles of a dead rattler remaining active for quite some time). I cut off the rattle, to add to my dad's collection of such keepsakes, and stuffed the still-writhing body into a burlap bag, thinking that someone might like a snakeskin for the ribbon band on their cowboy hat. Our group soon continued-on with our mapping project.

We finished our work and late that morning headed back to the vehicle. There, sprawled across the back seat we found Mohamed fast asleep with the back windows partly rolled down to give him air. Quietly the six of us surrounded the vehicle and I took the decapitated snake-enclosing burlap bag and lowered it though a window onto the floor next to him. Our field team then banged and banged on the front and sides of the vehicle to wake him and two of us vigorously pointed to the bag on the floor next to him. He opened it, peered in, saw this six-foot monster still twitching, and instantly tried to dive-out head-first though one of the half-open windows! We laughed, whooped and hollered in glee.

Why tell you this story? Well, rather than it being about our sleep-deprived laggard fellow student – who later went on to become the CEO and Chairman of the Global Environmental Facility, a Senior Fellow with the United Nations Foundation, and a world leader on global environmental issues – its focus is intended to be the rattlesnake. Yes, our field team defeated the potential peril, but we could only manage that because of the rattler's ectothermy and the time of day when it was encountered. We came across this snake early in the morning, before the environment had warmed up, so its movements were sluggish. Had we happened upon it at noon, after it had warmed to its surroundings, it would have been lighting-quick and it would have been immeasurably stupid for us to have confronted it. That's the deal about cold-blooded ectotherms. They depend mainly on external heat sources and their body temperature changes with the temperature of the environment. So, had this snake been a warm-blooded endotherm, using internally generated heat to maintain its body temperature, the whole bunch of us (except, or course, for Mohamed) would have been in deep trouble. The body temperature of an endothermic animal tends to stay steady regardless of the environment, and the resulting internally generated warmth speeds up muscles, neurons and all such body processes. That's why endothermy "won" the evolutionary competition.

Still, warm-bloodedness is not a panacea, primarily because it also means that endotherms require a lot of food – between five and twenty times more food than an ectotherm of similar same size – and thus require a food supply that is steady and reliable. During famines or in barren environments, endothermic animals may be less likely to survive than co-existing ectotherms which can get by with less energy. Similarly, under conditions of excessive cold or low physical activity many endotherms apply special mechanisms adapted specifically for heat production, mechanisms not needed by ectotherms. A good example is shivering (also called shuddering), a reflex reaction in warm-blooded animals when skeletal muscles begin to shake in small movements that create warmth by expending energy.

OK, we now understand the advantages of endothermy, but how did it evolve? Interestingly – and despite what you may have heard – a good case can be made that at least some meat-eating dinosaurs (sauropod reptiles!) must have been endothermic since it's hard to imagine such an active lifestyle being fueled by a solely cold-blooded metabolism. Moreover, several types of dinosaurs as well as reptile-derived proto-mammal synapsids exhibited prominent, mostly backbone-supported structures that seem clearly to have played a role in thermoregulation. For example, consider the 17 distinctive back plates (called "scutes") of the late Jurassic to early Cretaceous (159-144 Ma) dinosaur *Stegosaurus*, large triangular bony plates laced throughout with lattice-like structures and blood vessels. Folded onto the back, the plates helped keep the dinosaur warm, but raised vertically they radiated body heat into the environment (**Fig. 7-17A**). As for the proto-mammal synapsids, consider the large back sail of the late Carboniferous to early Permian (303-272 Ma) *Edaphosaurus* that is packed with parallel blood vessels as are those of their sail-backed synapsid relatives (such as *Dimetrodon*, *Secodontosaurus*, *Ctenospondylus* and *Lanthasaurus*). When the sail was raised, body heat carried by the blood would radiate away, but when it was folded onto the synapsid's backside it would keep heat in (**Fig. 7-17B**). Similarly, as we will soon see, in some reptiles and ancestral mammals their scaly skin was replaced by feathers and these, like the hairs of furry mammals are effective light-weight heat-retainers. Thus, unlike the strictly ectothermic and endothermic animals alive today, some dinosaurs and early-evolved synapsids were not exactly cold- or warm-blooded – they were evidently somewhere in between, ectotherms on the evolutionary path toward endothermy.

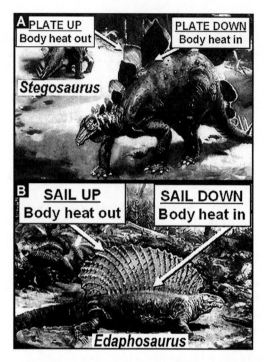

Fig. 7-17 (A) The Jurassic to Cretaceous plate-backed dinosaur *Stegosaurus* and **(B)** the Carboniferous to Permian sail-backed proto-mammal synapsid *Edaphosaurus*, illustrating the use of their blood vessel-packed back-structures to help regulate body heat.

The **Aves** (the Latin plural of *avis,* "bird") are birds, a taxonomic class that by including some 10,000 living species is the second most diverse group of living vertebrates besides bony fish. However, unlike fish, amphibians and reptiles, birds are fully evolved endotherms and all living examples have feathers, toothless beaked jaws and a strong yet lightweight skeleton. The class includes a great many bird species familiar to us: domesticated fowl – chickens, turkeys and ducks; small birds – hummingbirds, doves and chickadees; large birds – condors, penguins and flamingoes; birds of prey – eagles, hawks and falcons; and even flightless birds – ostriches, emus and cassowaries. We know birds well, mostly because they are obvious to us as they fly above in the sky but also because of amateur bird watchers ("birders") who by having the great fun watching and cataloging their comings and goings are a huge help to professional ornithologists.

The oldest known fossil bird is *Archaeopteryx lithographica* from the upper Jurassic (155 Ma) Solnhofen Limestone of southern Germany, a feathered flying (or at least gliding) reptile-like sauropod avian that exhibits characteristics of both birds and reptiles **(Fig. 7-18)**, shared features that firmly indicate the close linkage of the two groups. *Archaeopteryx*, for example, had teeth whereas modern birds do not. Ever hear the expression "rare as hen's teeth?" It works simply because chickens do not have teeth, at least the run-of-the-mill hens and roosters we know that like many other living birds use their muscular gizzards, sometimes filled with small stones known as "gastroliths" to grind down their food. Surprisingly, however, modern birds do have the genes needed to make teeth – all the genetic instructions needed for them to grow teeth have survived to the present – it's just that in modern birds the genes are turned off, no longer needed, no longer operating.

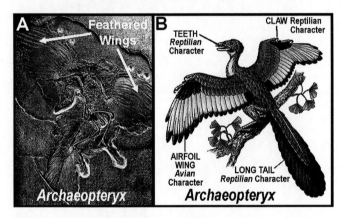

Fig. 7-18 (A) A fossil specimen of *Archaeopteryx,* the earliest known fossil bird from the Jurassic Solnhofen Limestone of Germany, and **(B)** a drawing showing its avian and reptilian characteristics.

Not every paleontologist or ornithologist is convinced, but the bulk of the evidence points to birds actually being avian reptilians. Yes, you read that correctly – birds being regarded as no more, no less than feathered flying dinosaur-like reptiles. In fact, quite a number of ancient small dinosaurs were demonstrably cloaked in lightweight heat retaining feathers, yet another adaptation on the path from ectothermy to endothermy. Except for the now long-known shared dinosaur-bird traits of *Archaeopteryx*, this "birds are really reptiles" notion is new, based on findings beginning in 1996 and the discovery of more than 50 new species of feathered dinosaurs in the fossil fields of China, chiefly in late Jurassic to late Cretaceous (160 to 65 Ma) deposits of northeastern China's Liaoning and Hebei Provinces **(Fig. 7-19)**. Given the shared dinosaurian and avian characteristics of *Archaeopteryx* and the new and rapidly growing body of supporting genetic and fossil evidence, birds have become increasingly pegged as simply "evolved feathered flying reptiles," the living members of this last surviving dinosaurian lineage being generally small due to the constraints of flight.

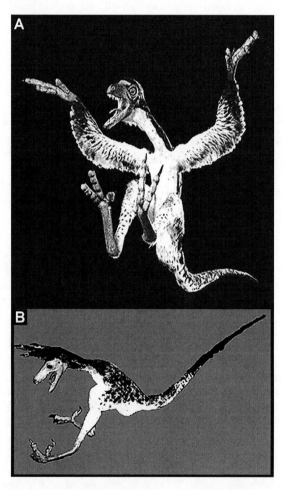

Fig. 7-19 (A, B) Fossil-based reconstructions of small feathered dinosaurs from the mid- to late-Mesozoic of northeastern China.

The rise of mammals – egg-laying monotremes to pouched marsupials to fetus-protecting placentals

The synapsid **Mammalia** (from Latin *mamma*, "breast"), the class of the Vertebrata that contains all animals that suckle their young, includes some 6,500 living species. All members of the class are endothermic amniotes and virtually all share the five defining characteristics of **(1) body hair**, which aids in insulation, **(2) mammary glands**, for production of milk, **(3)** a **cranial neocortex**, that part of the brain that enables complex thought, **(4) three middle ear bones**, for enhanced hearing sensitivity, and **(5) internal fertilization** of their embryos. The oldest putative fossil evidence of the mammalian lineage dates from the late Triassic (210 Ma) while the more widely accepted evidence is a bit younger, from the overlying Jurassic (178 Ma). In either case, the earliest mammals have been interpreted as being rodent-like insectivore multituberculates, a diverse and widespread group of early Mesozoic mammals and members of the only major branch of mammals to have become completely extinct.

The class Mammalia is itself divided into three groups, in their sequence of evolutionary development, the monotremes (such as the platypus), the marsupials (for example, the kangaroo) and the eutherians (all placental mammals from mice to man).

The key anatomical difference between **monotremes** and other mammals gives them their name derived from the Greek words *monos* ("single") plus *trema* ("hole"), thus meaning "single hole" and referring to the shared single duct (the cloaca) monotremes use for their excretory (both urinary and defecatory) and reproductive systems. Of this once abundant group – its history dating from the platypus-like fossil *Teinolophos* preserved in Cretaceous, 120-Ma-old sediments of the southeastern Australian Gippsland Basin – only the duck billed platypus and echidnas remain extant, the two types in total including a mere five species. Unlike other mammals, monotremes lay hard-shelled eggs and the egg-laying females lack nipples, secreting milk for the offspring through specialized hair follicles. Of the two types, the platypus is particularly weird, sporting a duck-like bill, webbed feet having venomous spurs, using electroreception as it swims and hunts, and glowing in the dark, together with opossums and flying squirrels being one of only three known biofluorescent mammals.

Marsupials are mammals in which the offspring are born under-developed, more complete development occurring within a pouch-like sac on the mother's belly after which the newborn becomes able to venture out on its own. The group includes such animals as kangaroos, wallabies, koalas, opossums and wombats, its name derived from the Greek word *mársippos* meaning "bag, pouch." Of the 335 species of marsupials living today, 235 occur in Australasia and 99 in Central and South America. The oldest fossil marsupial is *Sinodelphys*, known also as the "Chinese opossum," an extinct mammal dating from the early Cretaceous (125 Ma), whereas the oldest known more-or-less complete marsupial fossil, represented by dental, cranial and post-cranial remains, is *Djarthia* from the Paleogene (Eocene, 40 Ma) Tingamarra Local Fauna of Queensland, Australia. Clearly, therefore, the marsupial fauna for which Australia is justly famed has been in place for tens of millions of years.

The third group of mammals, the **Eutheria** (from Greek *eu-*, "good" or "right" and *thērion*, "beast," hence "true beasts") – more commonly known as "'placental mammals" – contains some 5,000 living species and is by far largest of the three mammalian groups. Here included are such diverse forms as whales, porpoises and dolphins; elephants, shrews, and armadillos; dogs, cats, sheep and cattle; and primates such as monkeys, chimpanzees and humans. Indeed, no doubt because we, too, are placental mammals, numerous now extinct members of the group have been enshrined as official state fossils – for example, the wooly mammoth, *Mammuthus*, the state fossil of Alaska, Nebraska, South Carolina and Vermont; and the saber-toothed cat, *Smilodon*, the state fossil of California. Like all mammals, the endothermic amniotic eutherians exhibit the five defining characteristics of the Mammalia but are distinguished from monotremes and marsupials by various traits of their feet, ankles, jaws and teeth, and by the absence of particular bones in their pubic region that allows eutherian placentals to expand their abdomens during pregnancy. The oldest-known fossil member of the group is *Juramaia sinensis* from early Cretaceous (161 Ma) sediments of China, the oldest specimen exhibiting both a skull and skeleton occurring in the mid-Cretaceous (125 Ma) Yixian Formation of northeastern China.

As we all now well know, the preservation of a fossil is a rare event, the deceased bodies decaying and their organic matter being recycled by other organisms back into the biosphere – a never-ending process along the path "from dust to dust." The same is true of mammals. Imagine an antelope on the plains of Africa set upon by a pack of lions. Its flesh would be devoured, leaving only its bones. But the rains would come and the bits and pieces of bones still remaining would be washed into a river and from there on to the sea. Given time, the mélange of bones, too, would be broken down and, like the flesh, recycled into the living world. But teeth are particularly robust, and being more resistant to such processes they tend to be longer lasting … think of sharks' teeth, for example **(Fig. 7-12)**. Thus, vertebrate paleontologists routinely can identify fossil species based only on their teeth – an easy matter for fossil mammoths and mastodons that even identifies their differing food sources **(Fig. 7-20)** – and experienced experts can give an accurate genus and species name of a great many fossil chordates based solely on a single or a few fossil teeth. Amazing stuff!

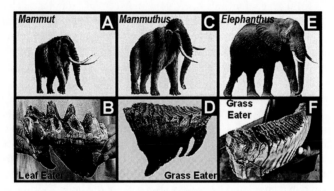

Fig. 7-20 Fossil and modern elephants and their teeth, **(A)** the extinct mastodon *Mammut*, **(B)** shown by its teeth to have been a leaf eater; **(C)** the extinct mammoth *Mammuthus*, **(D)** a grass eater; and **(E)** the living elephant *Elephantus*, **(F)** a grass eater.

How animals changed the planet – an overview

As we have seen in the last four chapters, animal life did indeed change the world. Here, then, is a synopsis, a ten-point summary of the history of animal life in the seas, skies and land.

(1) Given that all animals and plants are aerobic sexually reproducing eukaryotes it is not surprising that the precursor evolutionary stages leading to the emergence of animal-life were the same as those leading the rise of plants, namely the early Precambrian origin of oxygen-producing photosynthesis about three billion years ago, the mid-Precambrian increase in environmental oxygen that enabled the development of eukaryotic nucleated oxygen-dependent cells about two billion years ago, and the advent of sexual reproduction about one billion years ago **(Chapter 1)**.

(2) In the animal lineage, this sequence was then followed by the appearance of the soft bodied animals of Glaessner's pre-trilobite Precambrian (650 Ma) Ediacaran Fauna and, later, by the diverse hard-

shelled invertebrate animals of Walcott's 510 Ma Cambrian Explosion of Life **(Chapter 4)**. The progression began with Ediacaran colony-like aggregates of coenocyte-filled poriferan sponges and radially symmetrical two- tissue-layered jellyfish which by the mid-Cambrian had evolved to produce triple-tissue-layered bilaterian flatworms, early-evolved acoelomates in which the space between their tubular central gut and outer enclosing body-wall was completely filled with cells **(Chapter 5)**.

(3) The coelom, an open void between the tubular gut and the outside body-wall, also evolved by the mid-Cambrian and the animal tree of life then divided into two great branches, one, the protostome coelomate lineage **(Chapter 6)** leading through an array of invertebrate animals to insects, the other, the deuterostome coelomate lineage, extending from echinoderms to vertebrate chordates **(Chapter 7)**.

(4) The protostome branch, that in which the first hole in the embryonic blastula becomes the mouth of the adult animal, wended its way from lophophorates (brachiopod "arm foots" and bryozoan "moss animals," **Fig. 6-3**), to shelly mollusks (bivalve pelecypod "axe foots," gastropod "stomach foots," and cephalopod "head foots," **Fig. 6-4**), to segmented soft-bodied annelids and annelid-like onychophoran velvet worms having annulated legs **(Fig. 6-8)**, to hard-shelled jointed-legged arthropods (crabs, trilobites and their kin, **Figs. 6-9** to **6-11**) and a vast array of insects **(Figs. 6-15** to **6-17)**.

(5) In contrast, the deuterostome lineage – that in which the blastopore first hole in the embryonic blastula becomes the anus of the adult and the second opening becomes the mouth – traveled a path from diverse echinoderms (sea stars, brittle stars, sea urchins, sand dollars, and crinoids, **Figs. 7-3** to **7-6**) through the chordates from fish to amphibians to reptiles to birds and mammals.

(6) The earliest evolved such chordates were various types of fish, the most primitive of which, such as jawless fish (**Fig. 7-9**), platy-skinned fish and cartilaginous sharks lacked hard mineralized backbones. From these beginnings, bony fish evolved, a group having the greatest number of species of chordates living today and, thus, the numerically predominant members of the marine biota.

(7) Next came amphibians such as salamanders and frogs **(Fig. 7-13)**, four-legged vertebrates derived from the lobe-finned fish that used their muscular fins to slither across the mud from one drying pool to another **(Figs. 7-10** and **7-11)**. Although the adult amphibians lived on land they reproduced by soft gelatinous eggs ill equipped to withstand the rigors of the dry land surface. Amphibians were thus tied to living near water, stranded halfway between being land and water animals, a deficiency remedied by the evolutionary derivation of the amniote egg that was carried over to their descendants, the reptiles, birds and mammals.

(8) Like fish and amphibians, reptiles are cold-blooded ectotherms, deriving their body heat from the immediate environment. In comparison with warm-blooded endothermic birds and mammals that produce and regulate their body heat by internal metabolic processes, ectotherms have limited rapid muscle movement until the local environment, and thus their bodies, warm. To defeat this problem, various types of dinosaurian reptiles and synapsid stem mammals evolved special, mostly back-supported blood vessel-filled bony structures to keep heat in or to radiate it out **(Fig. 7-17)** whereas others developed scale-derived feathers **(Figs. 7-18 and 7-19)**, a lightweight form of insulation. Clearly, dinosaurian reptiles and synapsid proto-mammals were on the path toward endothermy, a process that came to fruition with the evolution of birds, more and more commonly regarded as "evolved feathered flying reptilians."

(9) The final phase in the deuterostome evolutionary progression was the development of chordate vertebrate mammals. All of these being endothermic amniote animals – all terrifically adept at protecting and reproducing their stock – the mammalian lineage began with monotremes, mammals that like their reptilian ancestors laid hard-shelled eggs. From this beginning, the progression continued though marsupials, mammals that produce immature offspring that mature in the mother's belly-pouch, to placental mammals like humans in which the fetus matures in the mother's abdomen. From this evolutionary past, placental mammals have come to dominant the land with some being successful in the seas (whales and dolphins, for example) and others, such as bats, flocking the skies.

(10) Like the most ancient groups of plants, few of the early-evolving animals have survived to the present. In comparison with ancient biotas, hardly any jawless fish or lobe-finned fish have lived-on to today; large amphibians and giant dinosaurian reptiles are also now long gone; and monotreme and marsupial animals are few and far between. Once dominant, once rulers of their world, almost all long ago met their demise. So it is in evolution, both for plants and animals: *"One day you can be the 'cock of the walk,' the next day a feather duster."*

Taken together, the history of Phanerozoic animal life can be divided into three great stages: The Paleozoic "Age of Invertebrate Animals and Fish," the Mesozoic "Age of Dinosaurs," and the Cenozoic "Age of Mammals." In the chapter to follow, we'll have a look at this evolutionary progression and see how it and the evolution of land plants fit together hand-in-glove.

CHAPTER 8

TAKE-HOME LESSONS ABOUT THE HISTORY OF LIFE

How is science done?

In 1962, philosopher and historian of science Thomas Samuel Kuhn (1922-1996) published a classic volume on how science advances entitled *The Structure of Scientific Revolutions,* a volume that in the early 1970's was required reading in no fewer than 14 UCLA courses in philosophy, the history of science, and sociology. Here is the story of Tom Kuhn and how his volume has had such an enormous impact on understanding of the workings of science.

As a graduate student in the mid-1960's, I, like many others read Kuhn's opus. Years later I came to know him personally, the two of us being members of the American Philosophical Society where each year we would seek each other out to chat, our friendship spurred at least in part because of our common background as members of Harvard's Society of Fellows when we each were graduate students. Born in Cincinnati, Ohio, into an educated family (his father having been an industrial engineer), Kuhn pursued his undergraduate and graduate studies in physics. Soon thereafter, given the freedom provided by his appointment as a Harvard Junior Fellow, he switched his focus to the philosophy and history of science. His classic 1962 volume was an offshoot of his earlier, 1957 book on the Copernican Revolution and the 1514 claim of the Renaissance-era mathematician and astronomer Nicolaus Copernicus (1473-1543; **Fig. 8-1A, B**) that the Sun, rather than the Earth is the center of the Solar System. Notably, Copernicus properly attributed this concept to having originated some 18 centuries earlier from the studies of the Greek astronomer and mathematician Aristarchus of Samos (310-230 BC; **Fig. 8-1C, D**).

Fig. 8-1 (A) Nicolaus Copernicus and **(B)** a drawing from his 1514 volume *Commentariolus* illustrating his Heliocentric Paradigm that placed the Sun, rather than the Earth, at the center of the Solar System. **(C)** The Greek astronomer Aristarchus of Samos and **(D)** an image from his 270 BC volume illustrating the spatial relations between the Sun and the encircling Earth and Moon, the concept resurrected by Copernicus 18 centuries later.

During our numerous conversations, Kuhn complained to me that his volume had been "all too often misread and misinterpreted," primarily by philosophers and logicians who had assumed him to imply that facts are malleable, changeable over time. He didn't mean that at all, firm facts being by definition immutable. Rather, his point was that the ***interpretation*** of the available evidence, the way that established facts fit or do not fit a given concept was variable, the overriding concept dictating the relevance or lack thereof of established knowledge. Unhappily, Kuhn harbored this distress over the misinterpretation of his writings until his death.

On a more pleasant note, Kuhn also told me that one of his other great regrets was that he was born too late (by some four centuries!) to have had an opportunity to personally interview Copernicus. I asked him why that would have been so useful. His answer was simple and direct: "Because I wanted to know about Copernicus' youth and whether his questioning of 'accepted dogma' was inborn or whether it had been acquired over his lifetime." In Kuhnian phraseology, by discovering the misplacement of the Earth relative to the Sun Copernicus had identified a fundamental "Anomaly" in the then-accepted "Paradigm," the trusted dogma that the Earth was the center of the Solar System and, thus, of the Universe itself. Copernicus had uncovered the discrepancy by carrying out what Kuhn called problem-solving "Normal Science" and, based on the evidence thus amassed, Copernicus had then proposed a new paradigm, a "Paradigm Shift" that has long-since been thoroughly confirmed. What Kuhn still longed to know was whether Copernicus' brilliant insight was a product of "nature or nurture," that is, whether it was inborn, dictated by Copernicus' genetics, or was implanted by his parents during his youth and his surrounding society as he matured.

I, of course, knew of Kuhn's work but, surprisingly to me, he knew of mine as well. Kuhn's notion was that like Copernicus, I too had initiated a Paradigm Shift by showing that the documented fossil record, and thus the process of biological evolution extends some seven times farther into the geological past than had previously been imagined, the earlier long-established dogma asserting that such evidence was "unknown and unknowable." Moreover, like that of Copernicus, this new knowledge has withstood the test of time. Given Kuhn's frustration at being unable to interview Copernicus and other prominent paradigm-shifters, he quizzed me about my youth. He then discovered that, well, yes, I was problem-solver early on, not always satisfied with the status quo – and, at the time, I regarded this inquisitiveness as not at all unusual, in no way inappropriate. Repeatedly, he urged me to write an account of these early years, a first-person narrative that he asserted *"would be of great value to future philosophers, historians and psychologists of science."*

In 2019, some two decades after Kuhn's death, I finally followed his directive by writing *Life in Deep Time: Darwin's "Missing" Fossil Record, A Personal Account of Paradigm Changing Science* (CRC Press, 229 pp.). As the writing proceeded, however, I found Kuhn's requested first-person narrative to be uncomfortable – far too many "I," "me," and "we" phrases. So, I recast the book in the third-person singular as though I were being interviewed by a knowledgeable reporter (which, after all, is what Kuhn wanted and in my view a bit of an experiment to show a way around the problem to others who might similarly dislike self-serving "me-first" presentations). Like Copernicus, and Tom Kuhn too, I tried mightily in my writing to give credit to others, both contemporaries and those who came before, important because that is the way that science actually progresses, building on and expanding earlier established understanding of the world.

In addition, in writing this book I had the fun of remembering and recounting a few of the problem-solving episodes of my youth. One such example was the time on my aunt's Iowa farm when I – an inexperienced "city boy" from Columbus, Ohio – figured out how to run down, isolate and catch a chicken bare-handed to bring to the farmhouse for dinner (a truly rather difficult task, chickens being really, really quick on their feet). Another example was the winter afternoon on a Friday in grade school when I shimmed up a bathroom wall and unlatched a window so that I could crawl in the next morning to open the school where our basketball team could then practice. As egregious as this break-in was, for several weeks all went well, even after someone spread the word and we were joined in the gymnasium by 40 or 50 other rambunctious fifth- and sixth-graders who, like us, wanted to play with their pals and escape the snow. After more than two months of this unauthorized great fun, one Saturday morning a teacher showed up and peered in. The whole bunch of us immediately cleared-out, scattering to the exits like leaves swept up in a massive windstorm.

Let us now return to the "Kuhn Cycle" that Tom Kuhn claimed my work has illustrated. What is the Kuhn Cycle and how does it proceed? The Copernican Revolution exemplifies the heart of the cycle, a "Paradigm" to "Normal Science" to "Anomaly" to "Paradigm Shift" gradually unfolding process. At the same time, the cycle also incorporates important additional features, most notably the reactions of established scholars to a proposed new paradigm, a concept that flies in the face of what they have long assumed to be essentially unassailable. Understandably, the annoyance and skepticism thus engendered results in what Kuhn terms a "Crisis Phase," a period of disbelief, doubt and debate that not uncommonly leads to a "Scientific Revolution" pitting the orthodox elders against the younger adherents of the new-view synthesis. Depending on how this revolution turns out – a fact-driven yet inherently human and thus emotion-laden conflict – a "Paradigm Shift" may occur and a "New Paradigm" set in place.

Let's walk through the cycle step-by-step to better understand its components **(Fig. 8-2)**.

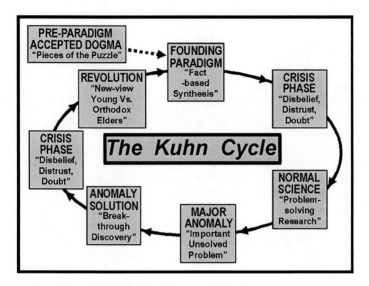

Fig. 8-2 The major stages of the Kuhn Cycle discussed in the text.

(1) Pre-accepted Dogmas

All Paradigm Shifts address previously accepted explanations of major aspects of the world around us. All thus involve the overturning of Pre-accepted Dogma, such prior understanding being primarily a function of the data available and the culture in which the previous paradigm has held sway. As a result, it is not uncommon for the new synthesis to be at odds with long-accepted religious beliefs that based on faith have been regarded as providing the answers to major questions about the untestable unknown. In short, the Kuhn Cycle begins with that which has come before, bits and pieces of the central question that the Founding Paradigm seeks to more accurately explain.

(2) Founding Paradigm

In virtually all cases, the new-view Founding Paradigm is a fact-based synthesis of the issue being addressed.

(3) Crisis Phase

Announcement of the Founding Paradigm engenders consternation in adherents of the previously accepted dogma, a Crisis Phase rejection of the new competing concept that can carry on for decades or even centuries.

(4) Normal Science

Undeterred, adherents of the new paradigm continue their problem-solving Normal Science.

(5) Major Anomaly
At some point -- either immediately or decades later – the accumulating evidence highlights an Anomaly, a major unsolved problem in the Founding Paradigm that had been overlooked or explained away.

(6) Anomaly Solution
As a result of ongoing Normal science, a "breakthrough" fact-based solution to this Anomaly is discovered.

(7) Crisis Phase
Announcement of the Anomaly Solution elicits a second Crisis Phase, in this case a period of disbelief, distrust and doubt among adherents of the Anomaly-containing paradigm, the solution of this previously unsolved problem being viewed as irrelevant or erroneous.

(8) Revolution
The ensuing and rather commonly acrimonious debate pits believers of the previous "old-school" paradigm against those who accept the new-view synthesis, a Kuhnian Scientific Revolution.

(9) Resolution
As a function of the quality of the evidence available and the prominence and influence of its promoters, a new world-view, a Paradigm Shift may ensue.

(10) Continuation
The cycle continues as an increasing body of fact-based information accumulates with each new line of evidence leading to a more complete and more fully supported understanding of reality.

Evolution – a test case of the Kuhn Cycle

To more fully examine the Kuhn Cycle – to see how it works and understand whether it truly describes the workings of science – we will now consider a "test case" not addressed in Kuhn's masterful opus, namely Darwinian evolution, its Darwin-noted Anomaly, the Anomaly Solution, and the resultant Paradigm Shift.

(1) Pre-accepted Dogmas
As with all newly founded paradigms, Darwin's evolution addressed a previously unquestioned explanation of a major aspect of human understanding, namely the history and development of life on Earth. Throughout all of Europe for the previous nearly two millennia – dating from the Roman Empire that stretched from Italy to the British Isles and across North Africa – the prevailing view was that God, as documented in the Christian Holy Bible had created all of life. Darwin,

like all people in all societies, was influenced by the culture around him and, thus, was brought up as an adherent of Biblical creationism.

A prime motivation of naturalists such as Darwin and his contemporaries was to elucidate by their studies of the natural world the wondrous workings of God. By no means did Darwin stand alone in this quest nor was he the first to uncover inklings of the naturally occurring, rather than a Biblically ordained unfolding of life's development. As noted in **Chapter 1**, the most influential to Darwin of his predecessors, those who before him had put together bits and pieces of his biological evolution-founding paradigm, were his grandfather, Erasmus Darwin, and Thomas Robert Malthus, Jean-Baptiste Lamarck and Baron Georges Cuvier. To this list should be added the ornithologist John Gould (1804-1881) who in 1836 recognized that the numerous varieties of finches that Darwin had collected in the Galápagos Islands west of Ecuador, known commonly today as "Darwin's Finches," were all derived from a single pair of founding parents, an important contributor to Darwin's concept of speciation through Natural Selection **(Fig. 8-3)**.

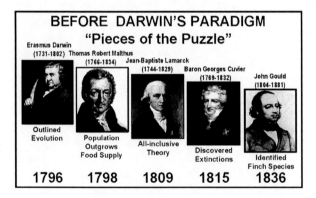

Fig. 8-3 Prime Pre-Paradigm contributors to Darwin's formulation of his all-encompassing concept of biological evolution by means of Natural Selection, from left to right, Erasmus Darwin (Charles Darwin's grandfather), T.R. Malthus, J-B Lamarck, G. Cuvier, and J. Gould.

In short, the precursors of Darwinian evolution – cultural and religious influences coupled with prior understanding of aspects of the Founding Paradigm – fit well with those prescribed by the Kuhn Cycle.

(2) Founding Paradigm

As stipulated by the Kuhn Cycle, Darwin's Founding Paradigm was a firmly based synthesis of the issue being considered, the great majority of the evidence coming from the collections he amassed during his service as the naturalist on the 1831-1836 world-spanning voyage of the *H.M.S. Beagle*. His collections were extensive – specimens of plants, animals, fossils, minerals, rocks and other "natural oddities" – virtually all of which were delivered for study to the natural history section of the British Museum (now London's Museum of Natural History).

By1844 Darwin had completed his opus, initially entitled "*An Abstract of an Essay on the Origin of Species and Varieties through Natural Selection.*" Then, as noted in **Chapter 1** – but contrary to the urging of his mentors, geologist Charles Lyell and botanist Joseph Dalton Hooker – he declined to permit his thesis to be published. Indeed, it was evidently not until the spring of 1858 that Darwin reconsidered the matter, prompted by his receipt of a letter from British naturalist Alfred Russel Wallace in Indonesia that enclosed a short manuscript presenting virtually the same broad synthesis as that in Darwin's opus. Shortly thereafter, on July 1, 1858 at a meeting of the Linnean Society of London, the Darwin-Wallace concepts were presented back-to-back, an event followed early the next year by the publication Darwin's manuscript by then bearing its revised title "*On the Origin of Species by Means of Natural Selection, or the Preservation of Favored Races in the Struggle for Life*". Clearly, Darwin's Founding Paradigm fits the Kuhn Cycle.

(3) Crisis Phase

As predicted by the Kuhn Cycle, Darwin's game-changing paradigm immediately engendered widespread resentment and rejection, primarily because of its reliance on a natural rather than a God-centered explanation of the unfolding of life. He was assailed and lampooned in the press, depicted as an "ape man" and caricatured as what one might call a sneering conniving "snake-oil salesman." The public outcry was fierce, so strident that Darwin's friend, commander of the *H.M.S. Beagle* Robert Fitz-Roy, having become unbearably upset for playing a role in Darwin's blasphemy of *"disproving the Bible,"* committed suicide.

Quite wisely, Darwin did his best to avoid the conflict, for the most part staying out of sight at Down House, his home at Kent in the near-London countryside. Moreover, Darwin regarded himself as an inept public speaker, unable and unwilling to present his views to an assembled crowd. But, of course, that was what was required for public education in the late-1800's pre-radio, pre-TV, pre-internet England. So, Darwin deferred this task to his spokesman, English biologist and anthropologist Thomas Henry Huxley (1825-1895) who became known as "Darwin's Bulldog." Though Huxley, like Darwin, was caricatured in the press, his talks were widely publicized and he was well aware of the consternation he precipitated, writing at the time that *"There is not the slightest doubt that, if a general council of the Church Scientific* [the scientific community] *had been held* [when *The Origin* appeared]*, we would have been condemned by an overwhelming majority."*

As the Kuhn Cycle predicts, the rejection of a new paradigm can carry on for decades, even centuries, for Darwinian evolution in some quarters lasting even to the present, more than 150 years after its presentation.

(4) Normal Science

As described by the Kuhn Cycle, adherents of Darwin's concept, undeterred by the Crisis Phase uproar, continued to carry out normal problem-solving science, in Kuhnian phraseology "articulating the paradigm." In this instance, an early but nonetheless excellent example is that of the German biologist Ernst Haeckel (1834-1919) who in 1866, only seven years after publication of Darwin's *Origin of Species,* used the newly unveiled paradigm to construct a "Tree of Life" of the living world, a compilation of Haeckel's well-founded observations and insights **(Fig. 8-4)**.

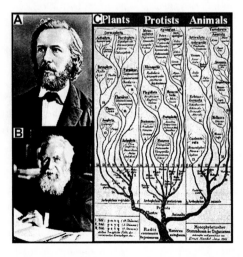

Fig. 8-4 (A, B) German biologist Ernst Haeckel and **(C)** Haeckel's Tree of Life, published in 1866, soon after publication of Darwin's *Origin of Species.*

Similarly, beginning in 1854 – a decade after Darwin's draft manuscript had been prepared but a few years before its publication – the Czech Augustinian monk Gregor Johann Mendel (1822-1884) conducted a series of breeding experiments on some 34 varieties of edible garden peas (*Pisum sativum*) that revealed the basic principles of heredity, the underlying keystone of Darwinian evolution. Mendel publicly presented his findings in1865, a scant six years after the publication of Darwin's *"Origin"* and nearly a century before discovery of the structure of DNA and the workings of genes. Thus, the focus of Mendel's experiments was how was to grow better pea plants, not on the genetic basis of their heritable characteristics. As a result, and even though he was the first to lay the foundation of the science of genetics – now known as "Mendelian inheritance" – the relevance of his findings to Darwinian evolution was ignored.

Some 50 years later, the picture markedly changed. In 1900, the independent findings of three scientists – the Dutch botanist and geneticist Hugo de Vries (1848-1935), the German botanist and geneticist Carl Erich Correns (1864-1933) and the Austrian agronomist Erich Tschermak von Seysenegg (1871-1962) – confirmed the results of Mendel's hybridization experiments. In Great Britain, biologist William Bateson (1861-1926), Master of St John's College at the University of Cambridge, became the leading proponent of Mendel's discovery of the laws of inheritance. Voilà, the science of genetics was born, with Bateson's work in particular shedding light on its relevance to Darwinian evolution.

As above, yet another salient stage of the Kuhn Cycle – the continued contribution of Kuhnian Normal Science following a Crisis Phase – has been confirmed for Darwin's Founding Paradigm.

(5) Major Anomaly

According to the Kuhn Cycle, at some point in the unfolding process – either immediately or decades later – the accumulated evidence may highlight a Major Anomaly, a seemingly crucial unsolved problem in the new world-view synthesis. For Darwin's Founding Paradigm, its principal glaring Anomaly was evident from the beginning. As Darwin wrote in *The Origin of Species,* *"If the theory* [of evolution] *be true, it is indisputable that before the lowest Cambrian stratum was deposited ... the world swarmed with living creatures.* [Yet] *why we do not find rich fossiliferous deposits belonging to these assumed earliest periods ... I can give no satisfactory answer. The case at present must remain inexplicable; and may be truly urged as a valid argument against the views here entertained."* And, so it was, trumpeted in the religious Creationist literature until the late-1970's, well more than a century after Darwin had broached the problem and more than a decade after the Anomaly Solution had been discovered.

Darwin's Founding Paradigm was based largely on two sets of evidence, one from the modern world, the specimens of plants and animals he collected during the voyage of the *H.M.S. Beagle,* the other far more ancient, the fossil record known particularly well in the British Isles due to the pioneering work of William Smith and his "Principle of Faunal Succession" **(Chapter 1)**. But in mid-1800, no fossils, no evidence of life whatever was known from pre-trilobite, pre-Cambrian rocks. To Darwin, this made no sense, and though he tried to explain it away, suggesting that early life may have been too small and fragile to be preserved, or that all the ancient fossil-containing rocks had been lost to erosion, none of his explanations held up. Thus, over the following 100 years this "missing" early fossil record came to be regarded as *"unknown and unknowable,"* regarded by some as *"the greatest unsolved problem in all of natural science."*

(6) Anomaly Solution

As is summarized in the Preface of this volume, even as a second-year college student this "missing" fossil record of the evolutionary process made no sense to me, either. So, I set out to solve the problem and, over time, I "marshaled the troops" worldwide. As a result, since the mid-1960's literally hundreds of pre-trilobite, pre-Phanerozoic (Precambrian) fossiliferous deposits have been discovered and the documented fossil record has been extended to 3,500 million years ago – a seven-fold increase from what was known only five decades ago. Moreover, the earliest known fossils, meshing well with modern gene-based phylogenetic evidence, show that primordial life evolved early, far and fast. The evidence is irrefutable – Darwin's dilemma about the *"inexplicable"* missing early record of life has been laid to rest.

The solution of this Anomaly in Darwin's Founding Paradigm well illustrates how the logic of science – and, in fact, that of everyday reasoning – actually works. Imagine that you take out your key to try to open the lock on your door. You turn the key to the left. The door does not open. 'Ha," you say to yourself, "the lock is broken." Then, rethinking the situation, you turn the key to the right … and the door opens. Or, imagine that you can't find your eyeglasses – you search and search and finally find them. Why is it that you failed when you first tried to open the door? Why do you always find your glasses in the last place you look? The answer is simple: you initially asked the wrong question!

In science, as in daily life, there are fundamentally only two ways that you can make a mistake when you are confronted by a problem. In some cases, it is because you did not properly understand the unsolved problem – in science, a failing easily remedied by reading and re-reading the relevant literature. More commonly, however, the problem is that you initially asked the wrong question. For the locked door, a better question would have been "do I need to turn the key to the left or, perhaps, to the right?" For the misplaced glasses, rather than approaching the problem by a random search, a targeted question – "where was I when I last put down the glasses?" – would have immediately presented the solution.

For the solution of Darwin's dilemma, the discovery of the so-called "missing" pre-Cambrian fossil record, the problem was much the same. Although those searching for the answer understood the problem, they failed to ask the right questions. Here, primarily because of the way Darwin and others phrased the problem, the paleontologists who sought its answer regarded it to be a simple matter of discovering fossils similar to those already known in older and older rocks. They thus searched the older strata for the same types of organisms having similar characteristics and preserved in the same way as those previously known. This, however, was the wrong set of questions, one that to the searchers seemed sensible chiefly because they were unfamiliar with what logically should have been expected in the older strata, namely simpler forms of life having more primitive characteristics and preserved in an entirely different manner from those then known from Cambrian and younger deposits.

How could this have been the case? Once again, the answer is easily understood, in this instance being a product of the education and prior experience of those seeking to solve Darwin's quandary. Virtually all the searchers were paleontologists, trained in geology departments and well-schooled experts on rocks, minerals and animal fossils. But these scholars had only limited knowledge of algae, fungi, or bacteria, more primitive earlier-evolved forms of life that fell under the purview of paleobotany, its practitioners trained in departments of biology, not geology. Moreover, the paleobotanical community had little interest in Darwin's "missing" fossil record, a major unsolved problem that pre-dated the rise of higher plants by more than 100 million years. Such differences were compounded by the differing modes of preservation of animals and plants, animal fossils being typically composed of preserved hard parts – their carapaces, shells, bones, teeth and the like – readily preservable structures absent from plants and their more primitive algal and bacterial precursors. Like plants, however, such earlier forms of life have robust cell walls and, thus, are best preserved by cellular permineralization ("petrifaction"), a mode of fossilization not exhibited by animals, the cells of which are enclosed by flexible membranes rather preservable cell walls. In retrospect, the anomaly might well have been solved far earlier had paleobotanists been interested in the problem – but they were not.

As is outlined in the Preface to this volume, the problem thus remained unsolved until 1965, a full century after it had been highlighted in Darwin's *Origin*, when – as the Kuhn Cycle indicates – its breakthrough solution was discovered as a result of Normal (problem-solving) Science in the form of two pre-trilobite pre-Cambrian ancient deposits, one in Canada and the other in Australia, both of which contained diverse cellularly preserved assemblages of readily identifiable fossil microorganisms.

(7) Crisis Phase

Immediately upon its publication in the spring of 1965, the first of these reports, that of the 1,900 Ma Gunflint microbiota of southern Canada co-authored by E.S. Barghoorn and S.A. Tyler, was met by the Kuhn Cycle-predicted flurry of disbelief, distrust and doubt, a Crisis Phase centered on the nature of the fossils reported and the their surprisingly great age. Numerous scientists expressed skepticism, both publicly and privately: *"The supposed fossils are unlike any I have ever seen, most likely they are tiny mineral grains"* ... *"they are far, far too old to be believable"* ... *"they don't make sense, there must be some terrible mistake."*

It is important here to remind ourselves that this was 1965 – more than half-a-century ago – and that this report flew in the face of the then-accepted dogma, the paradigm universally accepted by the scientific community expressed only two decades earlier by world-leading paleobotanist A.C. Seward's 1931 stinging criticism of C.D. Walcott's claim of fossil Precambrian bacteria **(Chapter 4)** that *"we can hardly expect to find in Pre-Cambrian rocks any actual proof of the existence of bacteria."* And there is little doubt that it was also influenced by the writings of the world-renowned paleontological guru George Gaylord Simpson (1902-1984) who in his classic 1944 volume *Tempo and Mode in Evolution* argued that the 500 million-year long "evolutionary distance" between modern man and Cambrian trilobites must be about the same as that between trilobites and amoebae (thought then to be first forms of life), from which Simpson deduced that the origin of life dated from only about 1,000 Ma ago.

(8) Revolution

What the paradigm-adherent naysayers did not know was that Barghoorn had yet a second arrow in his quiver, a similar but biologically more diverse and far-better-preserved microbiota of pre-trilobite, pre-Cambrian ancient microbes, that of the 850 Ma Bitter Springs cherts of central Australia. This find, reported by Barghoorn and me only a few months after the Gunflint paper, was the clincher – a wonderfully well-preserved assemblage of microscopic fossils preserved in much the same way as the Gunflint fossils but from a far younger deposit on another continent that included many fossil microbes easily relatable to those living today. Given this second report, the naysayers' palaver largely ceased and the Kuhnian Revolution Phase was thankfully short-lived.

(9) Resolution
.   Once again, the Kuhn Cycle correctly predicted the outcome of this game-changing view of Darwinian evolution. In the decade following the two breakthrough reports, as a result of Kuhnian Normal Science more than a dozen similar microbiotas were discovered worldwide, upped at present to many hundreds, and the new fact-based synthesis, a Paradigm Shift to the new view that life's history extends far earlier into the geological past than anyone had previously imagined, became widely accepted.

(10) Continuation

According to the Kuhn Cycle, science never rests on its laurels, its practioners probing incessantly to establish the scope and define the limits of the accepted world-view. It is thus germane to inquire how the new Paradigm has fared over the ensuing decades. Here the story gets even more interesting, Darwinian evolution having been tested time and time again as it has invariably held up to this repeated scrutiny.

How and why has that occurred? The answer hinges on what I refer to as the "Cascade of Evidence Test," the concept that "if this is true, then this must also be true, then this, and this, and this," an example of the workings of Kuhnian Normal Science. Indeed, science produces such test-and-test-again evidentiary cascades all the time, each affirmation of an independent line of evidence supporting the likelihood that the Paradigm is a correct description of reality.

Here is how such a Cascade of Evidence has substantiated the veracity of Darwinian evolution **(Fig. 8-5)**. The primary set of facts comes from the **fossil record**, a much longer and far more complete history of life than that known to Darwin as he prepared the *Origin*, a history that documents biological evolution from simple to complex, from single- to many-celled, from microscopic to megascopic, from microbes to plants and animals **(Chapters 2** through **7)**. However, for the fossil record to reflect Darwin's "Natural Selection" of organisms best adapted to their environment, the scenario must be consistent with evidence of the environment as well – as indeed it is, evidenced for example by the anoxic to oxygen-rich transition of **Earth's atmosphere** and life's adaptation to this major environmental change **(Chapter 1)**. Such evidence and many other relevant tests come from the geological sciences, all consistent with and supportive of Darwinian evolution.

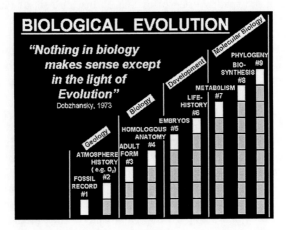

Fig. 8-5 The Cascade of Evidence Test applied to Darwinian evolution discussed in the text.

A second set of tests comes from the field of organismal biology, in particular the similarities and differences of the **adult forms** of life in the modern world. Thus, we humans like all other animals are vastly different from trees, shrubs and other members of the Plant Kingdom because we heterotrophs are not evolutionary closely related to such chlorophyllous photosynthesizing autotrophs **(Chapter 2)**. On the other hand, humans are particularly similar to chimpanzees. gorillas and other primates – both genetically and in social structure **(Chapter 1)** – and to a lesser degree to all other chordate animals as well, evidenced, for example, by the **homologous anatomy** of the limbs of humans, dogs, birds and whales **(Fig. 1-6)**. Countless other examples could be cited and the evidence is clear. Darwin's Paradigm holds the key, the basic fundamental concept that links together all life on Earth.

These first two sets of evidence, from geology and organismal biology, are supported by yet a third, that coming from developmental biology, the **embryos** and life histories of living organisms. As first shown by Ernst Haeckel in 1866 **(Chapter 1)**, the embryonic development of animals spanning the range from fish to salamanders to turtles to evolutionary more advanced vertebrates are all basically similar, diverging one from another only during the later stages of their early development. Similarly, the **life histories** of extant organisms – particularly, the universality among all higher forms of life of sexual rather than asexual reproduction – evidence their shared evolutionary heritage **(Chapter 1)**. As is true of the first two sets of evidence, this third set of facts and the test, retest and test again strategy of the Cascade of Evidence affirms the Paradigm of biological evolution.

Dating from 1953 and the discovery of the structure of gene-constituting DNA (deoxyribonucleic acid), a fourth set of evidence derived from the molecular biology of living systems has been added to the cascade. The first of the three components of this more recently added set of data, all based on the biochemistry of living systems, comes from studies of the **metabolism** of living systems, chemical processes that occur within a living organism in order to maintain life. Consider, for example, aerobic respiration, the oxygen-consuming process by which we and all other aerobes (Including plants) breathe. This metabolic process is composed of three steps, first glycolysis (the breakdown of the sugar glucose to form two smaller molecules of pyruvate), followed by the citric acid cycle in which the pyruvate is processed into products that are delivered to the third step in the sequence, the electron transport system, the only oxygen-using step in the process. In each of these steps the energy produced is cached in high-energy-storing molecules of ATP (adenosine triphosphate), but in the first two steps the energy-yield is low, a sum total of only 4 ATPs, whereas the third step, the O_2-requiring electron transport system yields a whopping 32 ATP molecules, eight times more than the other two steps combined.

Quite obviously, this high energy-yield has resulted in the prevalence and success of aerobic respiration worldwide, not only in all plants and animals but in microbial aerobes as well. Still, a central question remains unanswered, namely why is it that the three steps in the metabolic sequence differ so greatly in their energy-yields? The answer lies in the evolutionary history of aerobic respiration. The first two steps, glycolysis and the citric acid cycle originated in very primitive microbes that neither produced nor consumed oxygen – glycolysis being the metabolic process of the earliest forms of life, feeders on the primordial soup, and the citric acid cycle invented by somewhat later-evolved primal anaerobic non-oxygen-producing photosynthetic bacteria. Darwin's evolution is based on what he termed the "Natural Selection" of those organisms better adapted than their competitors to the same environment at the same time. By that criterion, in both of these instances the microbes embodying these innovations represented the "best game in town," each having quite effectively solved the problem of surviving and thriving in competition with their cohorts.

Though to us, such glycolysis- and citric acid cycle-originating microorganisms seem exceedingly primitive their competitors were as well – as the old adage has it, "in the land of the blind, the one-eyed man is king!" Indeed, it was only with the subsequent evolution of oxygen-producing and -using cyanobacteria that aerobic respiration arose, and it was not until hundreds of millions of years later that environmental oxygen had sufficiently built up to permit aerobic nucleated eukaryotic cells to come to the fore **(Chapter 1)**. Here, oxygen was the key ingredient, its metabolic use being an "add-on" to earlier-evolved processes, the complete package then being passed down from one lineage to the next. Although this is only one of a great many such examples that might be noted, it is sufficient to show that the metabolism of living systems is a product of biological evolution and yet one more test in the Cascade of Evidence that the Darwinian Paradigm passes with flying colors.

The second of the molecular biology-based tests in the evidentiary cascade is that based on the **biosynthesis** of the molecular components necessary for life to live. Here, as above, we need only one of many examples to establish the point, in this case the biosynthesis of the mammalian sex-determining steroid hormones testosterone and estrogen, both of which, interestingly, share their early biosynthetic pathway with that of carotenoids, the fruit-coloring pigments of higher plants **(Fig. 8-6)**. Each of these pathways begins with the same small starting compound, acetyl coenzyme A. After a few biosynthetic steps that produce a larger molecule, the pathways diverge and in each – as in the metabolism of aerobic respiration – the use of oxygen is a later addition evidencing that it is an "add-on" to the earlier-evolved part of the sequence. Once again, this biosynthesis test in the Cascade of Evidence substantiates its evolutionary history and, once again, it illustrates the co-evolution of life and its environment.

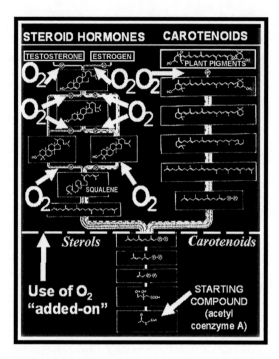

Fig. 8-6 The biosynthetic pathways of the mammalian sex-determining steroid hormones testosterone and estrogen (left) and of plant carotenoids (right) showing that that in both the use of oxygen is an "add-on" to earlier-evolved biosynthetic processes.

The third and last of this triad of molecular biology-based tests in the cascade is derived from organismal **phylogeny**, a gene-based concept that permits all extant living systems to be ordered into an all-encompassing Tree of Life. Following the late-1900's lead of Woese and Fox and their seminal studies establishing the three Domains of life **(Chapter 1)**, most such trees are based on rRNA (ribosomal ribonucleic acid), the evolutionary precursor and close cousin of gene-encoding DNA. Unlike DNA, which is housed in the gene-containing chromosomes of living systems, rRNA is found in the ribosomes of cells, tiny protein-(enzyme-) producing factories that are scattered across the cytoplasm in prodigious quantities (ranging from some 18,000 in a typical bacterium to 150,000 in each mammalian cell). Moreover, because all organisms, whether prokaryotic or eukaryotic, have ribosomes, and because ribosomes are largely composed of rRNA, this molecule can be used to document the relationships among everything alive today.

Although most such rRNA-based phylogenic trees focus on the evolutionary relations within a given biologic lineage or the derivation of one lineage from another, **Fig. 8-7** shows a compilation spanning all such data, a simplified total Tree of Life that encompasses all organisms alive today and all of the major biologic groups discussed in this book. As is abundantly evident, rRNA phylogenies well evidence and substantiate Darwin's Paradigm.

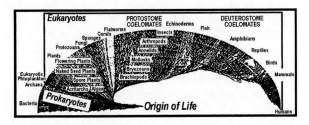

Fig. 8-7 A simplified rRNA Tree of Life that encompasses all organisms alive today and all of the major biologic groups discussed in this book.

In sum, it is important to note that each of the four major sets of fact-based data applied in this Cascade of Evidence test of Darwinian evolution – Geology, Organismal Biology, Developmental Biology and Molecular Biology **(Fig. 8-5)** – is separate from the others as are their respective components, the fossil record and atmosphere history; adult form and homologous anatomy; embryonic development and organismal life history; and metabolism, biosynthesis and molecular phylogeny. Thus, Darwin's Paradigm of biological evolution, now updated by a Paradigm Shift that solved its Major Anomaly, is thus confirmed by numerous discrete bodies of fact. Given that all of these varied lines of evidence mesh together to support the credence of Darwin's world-changing view of the history of life, there can be no doubt that biological evolution is a certainty, not some fake fanciful fable.

Sequential co-evolution of plants and animals

The genetically embedded two prime "laws" of all members of the living world are **(1)** to stay alive, the universal urge for self-preservation; and **(2)** to reproduce its kind, the urge to mate and produce viable offspring **(Chapter 1)**. Both are innate, both are coded in an organism's genes, and the success of each is relevant only in comparison with that organism's cohorts in a given environment at a given time. In other words, rather than needing to be "perfect" to survive and thrive, each form of life needs only to be a bit better than others competing for the same space and light (for plants) or food (for animals). That is the basis of Darwinian Natural Selection, the foundation of biological evolution and the reason that Earth-life has become increasingly varied, complex and successful in occupying the myriad nooks and crannies of the planet over its four-and-half-billion-year history.

In land plants, the "stay alive" dictum was answered by the evolution of improved stem anatomy and large megaphyllous leaves, innovations that permitted their bearers to grow taller and stouter, and to thus better compete in the side-by-side battle to soak up sunlight. In animals, it was responded to by the development of modified anatomical features that enabled newly evolved lineages to move onto the land surface and from there to the highlands, each stage providing access to previously untapped food sources. Similarly, both in plants and animals the "reproduce its kind" dictum was addressed by the evolution of new structures and new strategies, innovative changes that step-by-step improved the odds of reproductive success, each development better assuring that the offspring produced would survive and mature to reproductive age.

Nevertheless, aspects of both these prime dicta can be defeated, sometimes by chance events but more pervasively by the way an organism grows from its inception to adulthood. This is because in the life cycles of all organisms there is always a weakest link, that stage in their life history that is most susceptible to failure and, thus, a potential cause of the ultimate demise of the species. Moreover, in both plants and animals this stage invariably centers on reproductive processes, as we have seen for example in the evolution from spore-plants to naked seed plants to insect-pollinated flowering plants **(Chapters 2** and **3)**. The same is true of the evolution of chordate animals, a series of interrelated sequential events that allowed the more advanced lineages to spread into increasingly varied habitats. In land animals such innovations, whether anatomical (the development of limbs, **Chapter 7**) or physiological (the transition from ectothermy to endothermy, **Chapter 7**) were primarily spurred by the evolution of the hard-shelled amniote egg from its gelatinous unprotected amphibian precursor, an improved form of reproduction that because it provided better protection for the unborn embryo was carried forward and modified throughout all later-evolved vertebrate lineages.

Obviously, the evolutionary rise of plants and animals share similarities, both Kingdoms of Life having confronted the same or similar set of problems and both solving them by similar evolutionary adaptations. At first glance, that all seems sensible, perhaps even expected. After all, plants and animals are all eukaryotes, all have similar cells (plus or minus photosynthesizing chloroplasts), all breath oxygen, and all are derived from the same set of ancient primal precursors. All of these commonalities, however, are pre-plant, pre-animal innovations that reflect their shared unicellular eukaryotic ancestry. What about the structurally similar hard-shelled seeds of plants and the hard-shelled amniote eggs of animals? In both of the lineages these notably comparable structures not only house but also protect and provide nourishment – in essentially the same ways – to a developing embryo (**Fig. 8-8**). Still, if you stop to think about it, the striking similarity of the two types of hard-shelled eggs is actually quite surprising, seed plants and amniote animals being only distantly related (**Fig. 8-7**).

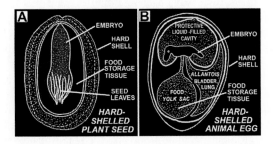

Fig. 8-8 Comparison of **(A)** a hard-shelled (gymnosperm) plant seed and **(B)** a hard-shelled (reptilian) animal egg.

Yes, evolution *does* build on that which came before, tweaking and adding additional small steps to previously established biochemical and developmental mechanisms. But this process can apply only to mechanisms that are readily accessible for modification, those encoded in an organism's genes, not to those occurring in some distantly related evolutionary lineage. Thus, and despite their obvious similarities, hard-shelled plant seeds and hard-shelled animal eggs arose absolutely independently. How could this be? To us, they look pretty similar and they do similar jobs. Isn't one a derivative of the other? No, not at all. Instead, they are simply similar solutions to similar problems – the prime advantage provided to each being the ability to invade previously unoccupied terrains. Indeed, the two types of protected eggs are innovations referred to by evolutionary biologists as illustrating "parallel evolution" (such as the similarity of "do-alike" analogous organs, for example a butterfly's wing and that of a bird) rather than being evolutionarily related homologues (as, for example, the wing bones of birds and bats; **Fig. 1-7**).

Perhaps even more interestingly, the evolutionary history of plants and animals well illustrates **sequential parallel evolution**, both groups solving essentially similar problems in similar ways with plants leading the charge into previously unoccupied habitats and animals trailing behind. This process is also properly referred to as **sequential co-evolution**, a similar phrase that highlights its two prime aspects, firstly its sequential nature – plants first, animals then following. And secondly, the plant-animal interactive bases of the evolutionary innovations, the presence of co-existing plant-devouring animals making it advantageous for plants to expand their range into animal-free settings and the animals, the "eaters" then following their food, the "eatees." Each lineage solved the problem in its own way, each building on and modifying its own previously evolved genetically encoded structures and mechanisms.

In common parlance, one might say that for plants the problem was "there's space over there, if we could 'figure out' how to use it." For animals, it was "there is food over there that we could use, if we could 'figure out' how to access it." In fact, however, neither of these statements makes much sense. For both plants and animals, the innovations were fundamental, deeply rooted in their genes, not minor modifications, not simply changes in behavior. Moreover, neither group "figured out" a way to address the problems, neither group analyzed the situation and devised a plan to solve it. That is not the way that evolution works. Rather, for each lineage a solution was at hand, one that came to the fore by the trail, error, and ultimate success of the workings of the evolutionary process. In other words, it was advantageous for plants to expand their range into unoccupied terrains where they could survive and thrive unaffected by plant-feeding animals, and it was similarly advantageous for animals to then expand their range to follow their food. That is what Darwin meant by "Natural Selection" – the best adapted win, the less adapted are winnowed from the population. That is the way that evolution works.

Given the foregoing, let's now retrace the sequence and see how the evolution of plants and animals fit together. Although the recent discovery in Argentina of fossilized liverworts, very simple land-inhabiting plants that lack stems and roots, suggests that plant-life invaded the land surface at least as early as the mid-Ordovician (472 Ma), the earliest known species of the spore-producing rhyniophyte *Cooksonia* **(Chapter 2)**, a true vascular land plant, was described in 2018 from mid-Silurian (432 Ma) deposits southwest of Prague, Czechoslovakia. Thus, full-blown land plants invaded the nearshore habitat at least this early, an advance that was then followed some 60 million years later by the evolution of plant-eating amphibians evidenced earliest by *Elginerpeton* **(Chapter 7)** found in late Devonian (368 Ma) rocks of Scotland. In short, spore-plants – tied to water by their reproductive processes – came first, followed by four-legged land animals – similarly linked to water by their soft-shelled easily dehydrated eggs. The evidence is clear. The evolution of plants, the "eatees," and animals, the "eaters," mesh together one after the other.

The other major sequential co-evolutionary plant-animal event is much the same. The earliest known seed plant, the first to exhibit the hard-shelled seeds that defeated the need for nearby water to reproduce, was *Elkinsia*, a seed-fern dating form the late Devonian (376-360 Ma) of West Virginia **(Chapter 3)**. The development of this new means of reproduction soon gave rise to plants that occupied both lowland dry terrains and highland locales. This, in turn, was followed some 55 million years later by the advent of reptiles with their hard-shelled eggs, the earliest now known being *Hylonomus* **(Chapter 7)** from the late Carboniferous (315 Ma) of New Brunswick, Canada. Once again, the "eaters" followed the "eatees."

It is notable that in both of these instances the time lag between the plant-offered opportunity and the animal response was quite long – 50 to 60 million years – but that is not surprising. After all, gene-based biological evolution is a slow step-by-step process of trial, error and ultimate success, not one that responds immediately to a new potentially life-changing opportunity and the prime reason why it is so difficult for us to observe evolutionary change in the world around us. What *is* a bit surprising is that paleobotanists and paleozoologists – experts in deciphering the fossil record – rarely highlight this history of the biota-changing sequential co-evolution of plants and animals, despite the fact that it is fundamental to understanding how the present-day living world came into being.

What might be the cause of this seeming oversight? There is no doubt that a prime aspect of this deficiency is the time lag between the plant-offered opportunity and the animal response, the sequential events being temporally so far removed one from another that they do not appear obviously interrelated. Probably more important, however, are the differences in the modes of preservation of fossil plants (commonly having intact cells and tissues) and fossil animals (primarily limited to their hard parts, their soft tissues having long-since decomposed), the preserved evidence thus prompting paleobotanists and paleontologists to ask differing sets of questions. Moreover, these differences are compounded by the physical separation and resulting differing training of the practitioners of these two subsets of paleobiology, in most universities paleobotanists being housed in the biological sciences, paleozoologists in the physical sciences. Finally, neither group seems particularly interested in the basic ecological plant-animal interactions of the floras and faunas they study, a central question being what plants were being eaten when, by what animals and why.

Taken together, such disparities are detrimental to paleobiologic science, hindering its ability to understand how the evolution of plants and animals are so inherently and deeply interconnected. After all, as this discussion has taught us, Nature is *not* compartmentalized! Indeed, as in many areas of inquiry, a more all-encompassing "holistic" approach – a matter of cataloguing the products of a process and then seeking to define and understand the underlying process itself – provides a more effective means to unravel and answer the question at hand.

Intelligence has ancient roots

As a final item of these "take-home lessons" about the history of life, we will now turn to the origin of intelligence, a human trait that we like to trumpet and one that is commonly imagined to set us apart from the rest of the living world. Actually, however, like virtually all other human traits, intelligence is derived from our evolutionary ancestors and, surprising as it may seem, has deep evolutionary roots. Here is the story.

To begin the discussion, we must first ask the question "what is intelligence?" Definitions vary, but the common dictionary definition we will use here is "the ability to interact with and by means of reasoning manipulate the environment effectively." Given this, how do we figure out its origin? Here a useful approach is to consider what logicians refer to as a "null hypothesis," an alternative hypothesis that states the logical opposite, and thus the negation of the proposition presented. In other words, for the roots of intelligence the problem becomes a matter of figuring out which kinds of life exhibit this trait, which do not, and ask why do they differ?

Let's begin with plants. Are they smart? For example, are ferns, or cycads, or pine trees intelligent? Some would claim that the answer is yes – after all, plant roots *do* disturb, "manipulate" the nearby soil; the pores (stomates) in plant leaves *do* quite sensibly open during daylight to insure the inflow of atmospheric gases and then close at night, to avoid evaporation of cellular water; and plants *do* multiply, proliferate and take over a landscape. But do these qualify as demonstrating that plants "interact with and by means of reasoning manipulate the environment effectively"? Not at all. Plants do not have brains, or muscles, or even a nervous system. More specifically, plant roots elongate and propagate to maximize the uptake of soil-water, growing downward in response to gravity due to the diffusion of a specific plant-growth hormone (auxin) from their root tips; the opening and closing of plant stomates is a result of cytoplasmic turgor pressure, not a result of reasoning; and plants expand their range into nearby land as they seek to survive, thrive and out-compete their challengers. All these are purely pre-programmed gene-based physical responses to their immediate surroundings, the plants reacting to and being governed by – rather than mindfully manipulating – their environment.

Simply put, plants sit in one place and vegetate. Why does that work? The basic requirements for life to stay alive, genetically encoded in the first "law" of living systems, are to have ready access to a source of CHON (carbon, hydrogen, oxygen and nitrogen, the chemical elements that make up living systems) and, secondly, a source of energy so that they can maintain their existence. The anchored-in-one-place life-style of plants meets these requirements, both being provided by the local environment, CHON from their immediate surroundings and energy from the sunlight they harness via photosynthesis to produce their life-sustaining glucose sugar.

Clearly, vegetating plants are not "intelligent." But what about sit-in-one-place animals, are they smart? Relevant examples are sedentary brachiopods, corals and stalked crinoids **(Chapters 5** and **6)**. All of these are basically "suspension-feeders," each using its appendages (lophophores for brachiopods, tentacles for corals, outspread arms for crinoids) to sponge-up suspended bits and pieces of food particles that are carried into their surroundings. Unlike plants, these and all other such non-mobile feeders (except sponges) have a brain, muscles and nervous system. Nevertheless, they too could not be called particularly "intelligent." Their basic needs are met, as are those of plants, by their anchored-down mode of life, in essence being "vegetating plant-like animals," obtaining their CHON and energy from the bits and pieces of food in their immediate setting.

Unlike such non-mobile plants and animals, organisms that can move about from one locale to another are decidedly different, and it is here that we first see the glimmerings of intelligence. Consider, for example, "lowly" flatworms, planarians, small worm-like bilaterian animals that use their cilia-coated undersurfaces to crawl across surfaces, even though they lack a coelomic cavity and therefore cannot burrow into an underlying substrate. The earliest-evolved members of the Animal Kingdom to exhibit a central nervous system and a well-organized brain housed at the front "encounter end" of their bodies, the frontal area of flatworms also includes two forward-looking light-detecting eyespots that enable them to move away from flashes of bright light. Moreover, laboratory experiments have shown that flatworms can learn to navigate a simple maze in search of food and that memories of such experiences are encoded in their brains, memory-banks from which they can later draw and thus need not relearn this previously mastered task **(Chapter 5)**.

Without doubt, flatworms exhibit at least a rudimentary form of intelligence. How do they use these primitive smarts, what benefit do they provide? In part, this trait is useful to find a mate and, thus, to up the odds of fulfilling the second gene-encoded "law" of life, to reproduce their kind. The primary function of such intelligence, however, is to search out food, the source of CHON and energy they need simply to stay alive. For flatworms, tiny ciliated "vacuum cleaners" sweeping up the food particles in their path, this would have been particularly important. Thus, by documenting the connection between intelligence and mobile feeding, "lowly" flatworms begin to answer the question of the how and why that intelligence evolved.

If that is really so – if intelligence arose fundamentally as a matter of self-preservation – shouldn't we see evidence of that in more advanced animal life? Of course, we should! And the evidence there is even more compelling. For example, consider the evolution of the feeding habits of coelomate (and thus potentially burrowing) worms. Yes, their fossil record is scanty – they are soft-bodied and consequently not readily preserved – yet "trace fossils," more formally known as ichnofossils, record their feeding behavior and evidence how they lived. Thus, as revealed by the burrows they produced, the fossil record documents their feeding habits that show an increasing ability over time to effectively interact with and manipulate their local environment. Primitive worm burrows from the latest Precambrian Ediacaran Geological Period cross through one another, the burrowers not recognizing that it makes little sense to re-mine food-containing mud after its resources have been exhausted. In contrast, burrows from the immediately overlying Cambrian Period show that later-evolved worms defeated this deficiency, each of their burrows mining only previously untapped portions of the substrate, a process later even better perfected by close-packed but non-overlapping "farming burrows" **(Fig. 8-9)**.

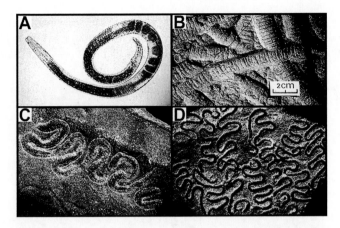

Fig. 8-9 (A) A modern coelomate worm, capable of burrowing, and **(B)** Precambrian (Ediacaran) worm burrows that cross through one another, **(C)** Cambrian worm burrows that do not overlap, and **(D)** younger Paleozoic "farming burrows" that mine the food-containing mud even more effectively.

But what about human intelligence, a trait we so love to tout? We are far different from worms! Does our intelligence share this common root, a primal search for food? To answer this question, we turn to the life around us. Consider birds, for example. Think for a moment about the "wise old owl." Why "wise"? Probably because of its oversized eyes, an adaptation that enables it to hunt at night, but maybe also because it is a lot smarter than we might have imagined. Or consider the bald eagle, the fearsome soaring symbol the United States (though Benjamin Franklin would have preferred the more placid North American turkey). Or consider a hawk that has hunted down a squirrel, a bird that has snared its prey by having the forethought to figure out where the squirrel will be, not where it was when it was first spotted. Clearly, all are intelligent, "able to interact with and by means of reasoning manipulate the environment effectively."

Ok, birds are smart. But we are neither worms nor birds, so how does this fit with us? Move up the Tree of Life and consider our mammalian cousins. Are dogs smart? Of course, even playful puppies and older hounds patiently waiting in line to leave their scent on a tree **(Fig. 8-10A, B)**. And horses **(Fig. 8-10C)**, of course, are also really bright as are, surprisingly, pigs **(Fig. 8-10D)**, both animals able to be trained to do stunts that they would normally avoid (with numerous studies showing that pigs are even easier to train than dogs). So it is with all our mammalian distant kin. But what about our nearest neighbors on the Tree of Life, primates to which we are most closely related – are they intelligent as well? Yes indeed! All such primates, whether monkeys or orangutans **(Fig. 8-11)**, or the chimpanzees and bonobos with which we share a huge preponderance of our genes **(Chapter 1)**, are certainly intelligent.

Fig. 8-10 (A) Playful puppies and **(B)** older hounds, all *Canis familiaris* and all subspecies derived from wolves (*Lupus*) domesticated by hunter-gatherers more than 15,000 years ago. **(C)** A highly trained intelligent horse (*Equus*). **(D)** A domestic pig (*Sus*), a social and notably intelligent animal.

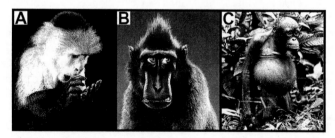

Fig. 8-11 (A) A Panamanian white-faced capuchin (*Cebus imitator*), a New World monkey native to the forests of Central America, the typical companion to a melodious organ grinder. **(B)** A Celebes crested macaque (*Macaca nigra*), an Old World monkey best known from the Indonesian island of Sulawesi. **(C)** An orangutan (*Pongo*), an intensively studied type of great ape native to Indonesia and Malaysia that uses a variety of modified natural objects, "tools," to construct elaborate sleeping nests and known to be particularly intelligent.

The case is made. What we call intelligence arose in mobile heterotrophs more than half-a-billion years ago, stimulated initially by the need for increasingly efficient food-intake, with intelligence being a truly valuable trait that has been passed on and gradually modified and improved over subsequent evolutionary history **(Fig. 8-12)**. But if that is so, when did the need for efficient food-intake actually originate? Here, again, the answer is readily apparent. This urge is not just primal but also primordial, extending back to the very beginnings of life and the feeders on the primordial soup. Like everything alive today, these earliest forms of life had to answer the need for CHON and energy simply to stay alive, to survive and thrive. It therefore follows that intelligence has exceedingly deep roots, extending even to the origin of life itself.

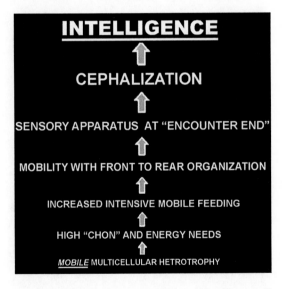

Fig. 8-12 A summary of the step-by-step evolution of intelligence discussed in the text.

Given all this, it seems obvious that the drive toward increasing intelligence in mobile animals is an inherent property of the Darwinian evolutionary process. And, as such it raises the intriguing possibility that this might also hold true on other worlds. We will investigate this question in the following and concluding chapter of this book by attempting to figure out whether intelligent aliens might actually exist and, if so and they were to visit our planet what they could learn about Earth-life and its exceedingly long history.

CHAPTER 9

A THOUGHT EXPERIMENT:
WHAT COULD "BLIND" ALIENS DISCOVER?

Thought Experiments

Thought experiments – basically, devices of the imagination – are made-up sometimes surprising and even humorous scenarios that can be used to illuminate aspects of the real world that otherwise may not seem so obvious. And though the creativity to put them together, the ability to effectively "think out of the box," is a relatively uncommon talent, it is much valued in science because it not infrequently provides a way to bring concepts together into a coherent whole that may previously have been grasped only piecemeal.

The physicist Albert Einstein (1879-1955; **Fig. 9-1A**) was a great proponent of thought experiments which he referred to in German as "*Gedankenexperiments,*" a trait that he enjoyed practicing even in his youth. As a 16-year-old, for example, he imagined chasing down a beam of light. A real-world experiment would have failed – light beams move at the speed of light – but this thought experiment planted in his mind the roots of his later-developed Theory of Special Relativity. This theory, based on his famous equation $E = mc^2$ showing that energy ("E") and mass ("m") are interchangeable as a function of velocity ("c" being the speed of light), codifies a relationship that explains how space and time are interlinked.

Fig. 9-1 (A) German-American physicist Albert Einstein, perhaps pondering yet another major unsolved problem about the universe. **(B)** Russian origin-of-life-biochemist Aleksandr Oparin, at the home of surrealist artist Salvador Dalí near Gerona Spain, 1973.

To set the scene for what now is common knowledge of yet another of Einstein's thought experiments, here is a bit of fairly recent history. In October 1957 the Soviet Union launched Sputnik, the first artificial satellite (after which Russia named its 2020 covid-19 vaccine). Only a month later, in November 1957, the Soviets sent up a small spacecraft housing the dog Laika, the first animal in space (though Laika died when the cabin overheated on the fourth orbit, an unhappy outcome that did not become public-knowledge until two decades later). Then, in April 1961, the Soviets sent up cosmonaut Yuri Gagarin, the first man in space, for a one-orbit 108-minute trip around the planet **(Fig. 9-2)**. He safely parachuted to Earth from an altitude of 23,000 ft (7 km) and landed about 10 minutes later near the town of Engels some 450 miles (724 km) southeast of Moscow. In May 1961, a mere month after this huge Soviet triumph, U.S. President John F. Kennedy announced to Congress that *"This country should commit itself to achieving the goal, before this decade is out, of landing a man on the Moon and returning him safely to the Earth."* This challenge was met in July and November 1969 by NASA's Apollo 11 and 12 missions to the Moon, safely returning to Earth six American astronauts accompanied by their caches of lunar rocks.

Fig. 9-2 Soviet cosmonaut Yuri Gagarin, on the cover of *Time Magazine*, April, 31, 1961.

Taken together, these events ushered in the Space Age, from which many of us have come to know yet another of Einstein's thought experiments, namely his "twins in space scenario." Imagine that you have a twin with the same genetics and born at almost the same time as you.

But the moment your twin is born, she or he gets placed in a spaceship and launched into space to travel through the universe at nearly the speed of light. According to Einstein's Theory of Special Relativity, you and your twin would age differently. Since time changes more slowly the closer an object gets to the speed of light, your incredibly fast-traveling twin would age more slowly than would you. So, when your twin returns to Earth, she or he might still be in diapers while you, approaching the later stages of life might be sorting out your retirement.

Thought experiments have also played a pivotal role in some of the science we have learned about in this book. For example, consider the origin-of-life "primordial soup" hypothesis **(Chapter 2)** put forth by the Russian biochemist A.I. Oparin in 1924 **(Fig. 9-1B)**. Like Einstein, Oparin came to this idea when he was a teenager, a 17-year-old soon to graduate from high school when he listened intently to a lecture at Moscow State University given by the Russian Darwinian-botanist K.A. Timiryazev. Yes, Oparin had the advantage of having earlier done some simple plant growth experiments – which led him to embrace Darwin's and Timiryazev's notions of biological evolution – but before this he had never considered how life itself emerged. He then made up a thought experiment to explain how life got started, a scenario that not only flew in the face of accepted dogma but for which he had no real-world supporting evidence. Indeed, it was not until 1953, three decades later and the breakthrough laboratory experimental results of the Miller-Urey non-biologic laboratory synthesis of amino acids that Oparin's primordial soup idea came to be widely accepted. (Interestingly, Oparin later opined that this experimental verification was to him unnecessary, in 1976 telling me *"a good theory, like this one, can stand alone."* It is also notable that only a year or so later Stanley Miller told me that neither he nor his mentor, Harold Urey, had even the slightest inkling of Oparin's ideas when they set out to do their 1953 experiments.)

Einstein's and Oparin's thought experiments, like many other such examples, show the value of this approach to problem solving. The ability to "think out of the box," to take what is known and then imaginatively expand that understanding into previously unexplored realms, can be a great impetus toward advancing human insight.

The "blind" aliens arrive

Here, then, is a thought experiment that illuminates aspects of life, environment, and the interrelated history of life and its environment, a central thrust of this book.

Benign intelligent aliens arrive in orbit around the Earth **(Fig. 9-3)**. During the trip to Earth from their home planet, a path-crossing asteroid collided with their spacecraft, the impact destroying their science instruments and making them all but inoperable. Virtually everything was lost – rendered totally unusable – including all instruments designed to measure the physical properties of the Earth, its place in the Solar System, its present-day environment and the distribution of land and sea, as well as its geological history and fossil record. Most other equipment was also affected, including a large suite of instruments designed to investigate in situ the attributes of Earth-life and its organism-encoded evidence of their evolutionary past.

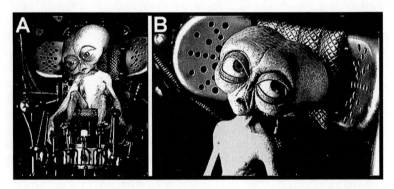

Fig. 9-3 Earth's benign inquisitive aliens **(A)** at the controls of their spacecraft and **(B)** shocked by the loss of virtually all of their planet-investigating equipment.

The only useable research equipment still remaining in workable condition was that designed to collect living organisms at systematically recorded locations, a menagerie comprising a representative sample of life on Earth – a veritable "Noah's Arc" representing all life on the planet – and to keep these organisms alive and return them to the aliens' home planet for future study. Thus, though the aliens are highly intelligent and far more advanced than humans, they have been rendered technologically almost totally "blind" by a chance encounter with rogue planetoid and until they return home will be unable to complete the task for which their mission had been designed.

Is an alien visitation plausible?

This imagined visit by a band of well-intentioned yet now scientifically "blind" space explorers is plenty fanciful. But to be useful, a good thought experiment has to be at least marginally plausible. So, does the scenario fill the bill? In a word, yes – and it is a lot more plausible than you might have imagined.

The human species, *Homo sapiens*, and all other bipedal primates of our lineage are of recent vintage, even "ancient" Neanderthals dating from only a few hundred thousand years ago. Thus, because the great majority of Earth's successful species have an average lifetime of about 10 million years, as evidenced in the fossil record, we hominids are in our mere infancy. Consider the future. Some 100,000 years from now, humans will no doubt look different than people do today (almost certainly having shed their current low-tech eyeglasses and those living in similar latitudes all having essentially the same skin coloration, a result of increased interbreeding and the latitude-dependent UV-protective effect of skin's melanin pigments). Societal structures might also differ from those today, perhaps resulting from the coalescence of all nations into a utopian all-world government and making this a planet of peace, harmony and mutual respect and cooperation.

All such notions, of course, are mere speculation … no one can accurately predict anything 100,000 years into the future! But in 10 *million* years – if we hominids survive that long without annihilating ourselves – there can be no doubt that we will be immeasurably different, by then hopefully having an evolved lower backbone anatomy to better offset the biologically "unnatural" bipedal stance of us genetically programmed quadrupeds, and old-age lumbar ailments and the use of wheelchairs and canes will no longer exist. And we can be sure that the visiting aliens would be at least a few million years in advance of Earth-life as we now know it, otherwise they could not have managed to actually visit and sample our planet.

If all that is true, why is there no scientifically accepted evidence of previous alien visitations? All have heard of UFOs (unidentified flying objects) or, in updated parlance, UAPs (unidentified aerial phenomena) including, for example, the Roswell New Mexico event of 1947. Though the Roswell episode was soon shown to be a due to a crashed conventional U.S. weather balloon, my Aunt June, a sheep-farmer on the outskirts of Roswell, was a firm believer as I learned during the summer I spent as a youth tending her flock. But, then – and despite being otherwise knowledgeable and intelligent – she was also an adherent of "reading palms" to discern a person's past and predict their future, and believed in phrenology as well, a matter of the practioner feeling the frontal lobes of a person's forehead and intuiting their intelligence and personality traits. Perhaps Aunt June was overly gullible.

In any case, time and again UFOs, claimed visitations, even mystical other-worldly reported contacts have been thoroughly, completely discounted, the sightings most commonly shown to be a result of atypical weather phenomena. And although a spate of recent reports remains unexplained, the unquestioned acceptance of such stories by some segments of the society, like the zealous support of various so-called "conspiracy theories," most probably reflects our all-too-human fascination with and fear of the uncontrollable unknown. A more logical explanation for the absence of compelling evidence of aliens is that though they are out there in the vastness of space, presumably roaming about on the lookout for fruitful sites to visit, the Earth and we humans are of little interest to them. After all, Earth is a cosmically unimportant "run-of-the-mill" planet. Moreover, Earth's intelligent advanced life is of only recent vintage, still young and immature as shown in part by the turmoil and unrest that has continuously plagued the human species for the past thousands of years. From the aliens' perspective, a preliminary overview survey of Earth life might be quite sufficient, a concerted study of humans having not yet risen to the status of being worth their time. And for the aliens, time might be a prime consideration. It might well be that Einstein was correct, that nothing can move faster than the speed of light and, therefore, that the effective use of the time allotted for their voyages is for the aliens of overriding concern, time itself being an unalterable precious commodity.

Let's now place the Earth and its life in its proper astronomical context. Our Solar System and all the planets within it date from about 4.5 billion years ago whereas the universe itself is some three times older, dating from nearly 14 billion years. From this it is easy to see that our Sun, a "Main Sequence" star having a lifetime of about 10 billion years, as well as the entire Solar System are already "middle-aged." When the Sun snuffs out billions of years into the future, the rest of the system will too. Further, our Solar System is just one of the great many solar systems that form our galaxy, the Milky Way. And the Milky Way is just a single galaxy in the universe, galactic systems estimated to number in the *hundreds of billions*. Current best estimates suggest that the Milky Way alone could contain upwards of 50 billion planets, some 500 million of which could be in their stars' habitable zones, locales where life could have originated and thrived. If you extrapolate those numbers to all the galaxies in the universe it's more than possible that the entire universe is teeming with literally billions upon billions of Earth-like planets, all capable of supporting life.

So far, our thought experiment scenario seems to make sense, all seems well and good. Humans, *Homo sapiens*, are in the infancy of their species-lifetime. There are literally billions of other planets, even in our own galaxy, and hundreds of millions where life could have taken root and thrived. And throughout the cosmos there are countless billions upon billions more such planets, some very much older than our planetary system. Given the numbers, it is obvious that a visitation by a civilization more advanced than ours seems plausible. But there is yet one last problem. Namely, if the speed of light is the prime limiting factor, could the aliens actually get here from their home planet and then manage to complete their voyage home with their collected Noah's Arc of Earth life?

To address this last question, let's consider what extra-solar system exploration we humans are potentially capable of carrying out. The closest stars to Earth are the three stars of the Alpha Centauri system. The two main stars are Alpha Centauri A and Alpha Centauri B which form a binary pair in which they revolve about each other. They and the third star of the system, Proxima Centauri, are an average of 4.3 light-years from Earth. Because Proxima Centauri is the closest to Earth, its planets represent the most sensible targets for a first interstellar mission. Moreover, in 2016 astronomers announced discovery of a rocky planet in the habitable zone of Proxima Centauri, a planet now known as "Proxima b" and a planet almost the same size as Earth, only 1.3 times as massive as our planet. Could Proxima b harbor life? We humans are inquisitive. We will want to find out.

Off we go? Well, not quite. How do we get there? How long would it take for a here-to-there and return-to-Earth expedition? The Alpha Centauri system is "only" 4.3 light-years from Earth. What does that mean? In particular, what is this "light-year" business? According to Einstein's Theory of Special Relativity, nothing in the universe can move faster than the unchangeable speed of light (186,000 miles per second, meaning that a "light-year," the distance light travels in one year is nearly 6 trillion miles). Thus, given Earth's elliptical orbit around the Sun, it takes on average 8.3 minutes for sunlight to arrive at Earth and it could get here neither earlier nor later.

More importantly, the speed of light is far, far faster than even of our most advanced rocket ships. Consider, for example, NASA's New Horizons spacecraft, among the fastest manmade objects ever launched from Earth and the first spacecraft ever to visit Pluto, the outermost member of our solar system. Launched from Earth in mid-January 2006 and traveling at its cruising speed of 36,000 miles per hour, New Horizons reached Pluto in mid-July 2015 – nine-and-a-half years later. Yet Pluto is a "mere" 3 *billion* miles from Earth whereas the Alpha Centauri system is about 25 *trillion* miles away, more than eight thousand times as far from Earth as Pluto and some three hundred thousand times the distance from the Earth to the Sun. So, if New Horizons were aimed toward the Alpha Centauri system it would take about 78,000 years to get there. In other words, the one-way trip would take some 240,000 human generations. This, in turn, would mean that to have reached Proxima Centauri now, today, the astronauts and their families would have to have boarded the spacecraft in 76,000 B.C. – way, way back when we humans were tribes of hunter-gatherers stalking mammoths and mastodons, collecting firewood, and foraging the wilds in search of fruits, berries and what other edibles could be found. And were these long-distant cousins to have landed on Proxima b today, they could not get back to Earth until the calendar year 80,000, some seventy-eight centuries into the future!

Nevertheless, even this last hang-up does not defeat our thought experiment. Given the age of the universe, the huge numbers of stars, galaxies and habitable planets, and the relative youth of planet Earth in comparison to a great many planets in the universe, the odds seem insurmountable that alien life much older and no doubt wiser and more capable than we exists. After all, Earthlings like us are neophytes while older more advanced civilizations, especially those capable of sending beings from that society to far-distant planets will have mastered feats that to us are all but unimaginable. In this regard and to set our minds further at ease about the plausibility of our scenario, it is useful to recall the 1962 dictum of futurist Arthur C. Clarke (1917–2008) that *"any sufficiently advanced technology is indistinguishable from magic."* Clarke was no doubt correct. Though we far-from-magical humans have had a tough time coping with even our most obvious and seemingly mundane societal problems, advanced aliens will have solved such difficulties – the longevity of their civilization proving the point.

So, let's keep to the thought experiment scenario and imagine that a team of benign intelligent aliens "magically" arrive at Earth, and even though the great majority of their instruments are "blind" they nevertheless collect a representative "Noah's Arc" of Earth-life and return this prize to their home planet for in-depth study. What could they then learn?

What could the aliens discover?

The aliens complete the trip and return home. Having been unable to complete their other assigned tasks they are accompanied only by their amassed "Noah's Arc" of Earth-life and the small volumes of the local environment obtained together with the life-forms collected at carefully recorded localities. What could they figure out about Earth-life? About Earth's physical properties and place in the Solar System? And what could they deduce about Earth-life's past evolutionary history?

In the following, we will consider these questions in turn – an abridged listing that represents only a smattering of the secrets that could actually be revealed by their cache.

First, what could the aliens conclude about the *basic overall characteristics of Earth-life*? In particular:

(1) Earth's biotic composition – its two cell types and three Domains of Life.
(2) Earth's biotic distribution – life's latitudinal variation from the equatorial to the polar regions of the planet.
(3) Earth's biotic structure – the dependence of "eaters" (heterotrophs) on co-existing "eatees" (autotrophs).

The following section then considers what the aliens could conclude about the *physical properties of planet Earth*. Taken in turn:

(4) Earth's current environment – its relative abundance of land and sea, its local and global average temperature, its atmospheric composition and pressure.
(5) Earth's global properties – its mass, shape, rate of rotation, and the current number of days and months in an Earthly year.
(6) Earth's astronomical setting – the existence of, and Earth's relation to the Sun and the Moon, the nature of the solar flux that reaches our planet, and the approximate age of the Sun and our encircling planet.

The concluding section then addresses what they could surmise about *Earth-life's past history,* specifically:

(7) How Earth's environment changed over time – day-length, Earth-Moon relations, atmospheric composition, ambient UV-flux, and surface temperature.

(8) How life began on Earth – from simple to complex, first primitive anaerobic primordial soup-derived heterotrophs and only later sunlight-requiring autotrophs.

(9) How life changed over time – from strictly anaerobic microbes to non-oxygen-producing autotrophs to oxygen-producers and -users; from single-celled eukaryotes to large multicellular plants and animals; from fish to amphibians to reptiles to birds and mammals; and the relative timing of these events.

Quite a list – and all this even without the telling evidence of fossils, the rock record, and measurement of geological time! It has taken humans literally centuries to pull all this together but there is little doubt that the aliens could find out all this and more – without the slightest inkling of what we Earthlings have already discovered – given their experience and advanced knowledge and technology. Still, if you stop and think about it, you will realize that this is not all that surprising. After all, it is abundantly clear that *all* life on Earth is well adapted, genetically honed to its particular local environment – otherwise it would not exist. The aliens would see this immediately, so a systematic perusal of their menagerie of Earth-life would easily move the most basic of their initial "unknowns" to their target list of "well established."

Basic characteristics of Earth-life

An initial survey of the aliens' cache would immediately uncover the basics of life on this planet, including:

(1) Earth's biotic composition – its two cell types, non-nucleated prokaryotes and nucleated eukaryotes, would be instantly obvious as would, with a bit of additional study, the similarities and differences between the two principal groups of prokaryotes. From this they could easily deduce the three-Domain, Archaea, Bacteria, Eucarya structure of Earth's biota **(Fig. 1-12)**. Indeed, given their advanced technology – for example, the easily conceivable availability to them of highly advanced biochemistry-evaluating high-precision electron microscopes that could simultaneously and more-or-less instantaneously obtain exceedingly high-magnification images and biochemical analyses of the structure and chemistry of cells in three dimensions – they might even discover additional categories of life.

(2) Earth's biotic distribution – life's latitudinal variation from equatorial to polar settings – would similarly be obvious from their collection in which the localities of their samples had been carefully recorded. Thus, for example, it would be a simple matter for them to see that the abundance of the different types of Earth-life varies systematically from a diverse, richly populated band across the central region of their map (for example, the Amazon rainforest) to its relatively depopulated upper and lower margins (walruses, polar bears, no trees, hardly any plants at one edge, and seals, penguins and not much else at the other). From this, it would then be an easy matter to discern:

(3) **Earth's biotic structure** – the dependence of "eaters" (heterotrophs) on co-existing "eatees" (autotrophs). The obvious co-occurring biotic density decrease of plants and animals from the mid-section of their map to its top and bottom would show that the diversity and abundance of plants and of animals track together. And from the experience gained by having kept the menagerie alive on the return voyage to their home planet, they would realize that plants are "light-eaters" whereas animals are "eaters of others." By coupling this pair of simple observations, the basic structure of the biosphere, the dependence of animals on co-existing plants would be immediately apparent.

Physical properties of Planet Earth

Similarly, it would be easy for the aliens to sort out **(4) Earth's current environment** – its relative abundance of land and sea, its local and global average temperature, and its atmospheric composition and pressure.

The aliens' plot of locality data for their amassed specimens would show that the diversity of life forms in their collection, both of heterotrophs and autotrophs, varies not only from the central region to their map's top and bottom, but also within areas of the map itself. Some areas, roughly 30% of their map, would have high concentrations of the various life forms that would diminish at the margins of the areas, whereas the remaining 70% of the map would be decidedly less populated. In those areas of the 70% far distant from the more populated 30%, diversity would drop to near zero, most prominently containing phytoplankton and, at depth, microbial Archaeans and sulfur-cycling Bacteria.

Combining these observations with the equally evident differences in the morphology of the inhabitants of the two major planetary segments – the autotrophs of the 30% being larger (trees, shrubs) and its most evident heterotrophs having four more-or-less flat-ended appendages (amphibians, reptiles, mammals), a situation in marked contrast to the smaller autotrophs (seaweeds, phytoplankton) and the disc-shaped (jellyfish) and tapered elongate heterotrophs (fish) of the 70% – they would deduce immediately that the life forms of the two regions were adapted to differing habitats. The differing morphologies and their samples of the local environments collected together with the organisms would hold the answer, the 70%ers inhabiting a more viscous fluid environment (the seas) and the 30%ers an environment that lacked such this fluid (the land areas).

Coupling this understanding of the distribution of life-forms with that gained from maintaining viable specimens on their return flight, they could then easily deduce Earth's local and planet-wide average temperature and its atmospheric composition. Their collection-maintaining temperature regime would yield the temperature data, with the optimal growth conditions of the vast majority of their specimens being in the 68° to 113°F (20° to 45°C) mesophilic temperature range whereas the rarity of extremophile high temperature- and high pressure-requiring Archaea in the collection would substantiate their conclusion.

Firm knowledge of Earth's atmospheric pressure would be only slightly trickier to obtain. For the 30% of their samples collected from land areas, their atmospheric pressure deductions would seem sensible for the gaseous envelope surrounding an Earth-sized planet, but although their deductions from these samples might therefore seem good enough, for the aliens "good enough" would not be good enough. So, to obtain confirmatory evidence they would check out the tallest organism in their collection, the California redwood (*Sequoia sempervirens*). Finding it to tower to a height of 380 feet (116 m) and by then knowing its local environment and the basic equation of autotrophic photosynthesis ($CO_2 + H_2O$ + sunlight $\rightarrow$ "CH_2O" + O_2; **Chapter 2**), they would realize that water, H_2O, was required to keep the redwood alive and that this water had to be transported from the ground, up the stem, to the uppermost leaves. But they would find that rather than being mechanically pushed up the tree's woody (xylem) core – plants having no nerve net, no muscles – it was actually being pulled up the narrow xylem tubes by capillary action, by droplets of water evaporating and being lost into the atmosphere at the top of the tree (a process that botanists refer to as "transpiration"). The height-limit of such autotrophs is thus a function of the overlying atmospheric pressure, an understanding that would reinforce the aliens' initial deduction.

From related studies of their "Noah's Arc," the aliens could also deduce **(5) Earth's global properties** – its mass, shape, rate of rotation, and the current number of days and months in an Earthly year. Earth's mass, the gravitational attraction experienced by Earth-life on the 30% of the planet where they are not supported by a surrounding viscous watery medium, would be easily calculated from the size of its largest inhabitants, African bush elephants (*Loxodonta africana*) up to 11 feet (3.3 m) tall and weighing as much as 6 tons (5,440 Kg). The spheroidal shape of the planet would be evidenced by the distribution of autotrophs in the sampled biota, particularly their marked decrease in abundance and types from the center of the map to its top and bottom margins. Because autotrophs live by light-requiring photosynthesis, the greater abundance of such autotrophs and co-occurring heterotrophs would imply the availability of more light toward the center of the planet and its decrease, its relative diminution toward the edges. A spheroidal planet would meet these requirements whereas a flat disc-like planet would not, with just such spheroidal planets being characteristic of planetary systems throughout the universe as the aliens would have seen countless times on their voyages though space.

The collected menagerie would also reveal Earth's rate of rotation as it spins daily on its axis, its current day-length recorded in what are called "circadian rhythms," a daily, at present 24-hour clock genetically embedded in virtually all animals (most mammals and birds sleeping at night while bats and owls fly about into the wee hours and then bide their time during daylight). The aliens could also discern the number of days in an Earth-month and Earth-year. Numerous near-shore marine animals such as corals and mollusks, for example, add regular layered increments each day to their hard calcite casings. Because these record the number of days in a month, the monthly high-tide additions to the layering being obviously thicker, and because their once-a-year (usually spring season) reproductive spawning cycle also leaves a telling trace, the aliens would find it an easy matter to count up the number of days in a month and the number of months in a year **(Fig. 9-4)**.

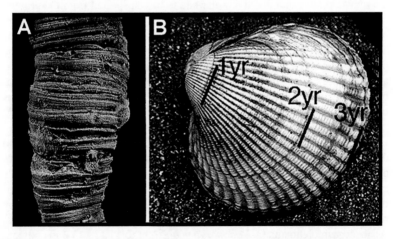

Fig. 9-4 Daily and monthly growth bands in **(A)** a coral and **(B)** a bivalve mollusk.

The validity of this simple technique to figure out the number of days and months in an Earthly yearly cycle was first shown in 1962 by Cornell paleontologist John W. Wells (1907-1994). Continued work has shown that it is applicable to corals and mollusks both modern and fossil. Wells' pioneering studies of fossil corals established that the number of days in Earth's year has decreased over geological time, a year in the Devonian Period 400 Ma ago being composed of 400 days not the current 365, with day-length in the Devonian being correspondingly shorter, about 22 hours, not the present 24. Mollusk shells from the Late Cretaceous, a closer-to-the present 70 Ma ago, show the same, year-length having decreased by then from 400 to 372 days and day-length having increased to from 22 to about 23.5 hours. Though the aliens would lack such fossil-based evidence, from studies of the living corals and mollusks in their menagerie they could nevertheless figure out the number of days and months in a current Earthly year.

If you've not previously heard about this day-month-year stuff, here's how the system operates. The total time it takes for Earth to complete its once-per-year orbit around the Sun is constant, not subject to change, and that is what defines the real-time (hours and seconds) length of an Earthly year. The Moon is kept in orbit by the gravitational force that the Earth exerts on it, but the Moon also exerts a gravitational pull on the Earth. On the "near" side of the Earth, the side facing the Moon, the gravitational force of the Moon pulls the ocean's waters toward it generating a "tidal bulge." This creates tidal friction between the two bodies that over time slowed the Moon's rotation such that it ultimately became tidally locked in one position – it stopped rotating leaving its same face, its same side forever pointed toward Earth. The slow migration of the Moon away from the Earth is also mainly due to the action of Earth's tides. The tidal bulge feeds a small amount of energy into the Moon, gradually pushing it into a higher orbit. In essence, this is similar to what a youngster feels on a playground roundabout, a simplified stripped-down version of a carnival carousel. The faster the roundabout spins the stronger the feeling of being slung outwards by the centrifugal force.

Thus, the Moon is being pushed ever so gradually away from Earth. This continuous, day-by-day, month-by-month increase of distance between the two bodies results in a corresponding slight alteration of the effects of their body-to-body gravitational interaction which, in turn, causes the rate of Earth's spin to correspondingly decrease and the length of an Earth day to therefore gradually increase. These interactive relationships are prescribed by the laws of physics, commonly phrased as being required "to maintain the invariant total angular momentum of a closed system." Although to non-physicists unacquainted with such rules this may seem a bit arcane – and although the Earth-Moon system is not actually "closed," the two bodies being pushed and pulled by the gravitational attraction of other components of the Solar System, primarily the Sun – an instructive way to understand the Earth-Moon system is to compare it to a spinning Olympic ice-dancer. As the Olympian begins to spin, arms far outstretched, the skater spins slowly. The dancer's arms are then gradually pulled ever closer to the body and the spin-rate correspondingly increases to a crowd-pleasing dazzling maximum **(Fig. 9-5)**. Then, to slow the spin so that the Olympian can move on to the remainder of the dance routine, the dancer does the reverse, gradually extending the arms and slowing to a stop. Aha! The Earth-Moon system is like a spinning Olympic ice-dancer!

Fig. 9-5 (A) As a spinning ice dancer contracts her arms ever-closer to her body, **(B)** her spin-rate correspondingly increases, a good example of the physicists' law of the "conservation of angular momentum."

At present, the "dance" of the Earth-Moon system is continuing to slow as the Moon continues to move away from the Earth, currently at a rate of about 1.5 inches (3.8 cm) per year as shown by measurement of the time required for laser light to be bounced back to Earth by arrays of laser-reflecting mirrors emplaced on the Moon by NASA's Apollo 11, 14 and 15 missions and two Soviet Lunokhod missions. Moreover, this gradual recession of the Moon from the Earth has been occurring for literally hundreds of millions of years, as the fossil data discussed above well illustrate.

Thus, from studies of the living shelly invertebrate animals in their collection, the aliens could discover the number of days in a month and the number of months in a year.

But what about **(6) Earth's astronomical setting** – the existence of, and Earth's relation to, the Sun and the Moon, the nature of the solar flux that reaches our planet, and the approximate age of the Sun and its encircling planet Earth? Could they figure out all this, too?

Again, this would not be a difficult task. The existence of our local star, the Sun, would be evident from the dependence of Earth's autotrophs on light, without which they and the great majority of the entire biota could not exist. The presence of Earth's satellite, the Moon, would be shown by its monthly tidal influence on corals and mollusks, as discussed above. And the nature of the solar flux reaching the surface of our planet would be evident from the aliens' understanding of Earth's atmospheric composition – a portion of which deflects or absorbs solar rays – and the visible light-absorbing qualities of chlorophyll, the green pigment that colors the leaves of the trees, shrubs, grasses and even the "pond scum" (cyanobacteria) with which we are familiar.

To reinforce this solar flux-indicating finding they would then study the absorbance characteristics of the pigments of the anaerobic sulfur bacteria and purple bacteria that in modern cyanobacterial mat-communities populate a thin zone immediately below the chlorophyll-bearing visible light-absorbers **(Fig. 9-6)**. From this "natural experiment," enabling them to measure the wavelengths of light left over after those absorbed by chlorophyll had been subtracted by their use by the overlying microbes, they would immediately realize that red and near-infrared light also penetrate to Earth's surface. From this understanding of the solar flux they would know that the emitter of this light-spectrum was a run-of-the-mill Main Sequence star – not a late-phase Red Giant or an even-later-phase White Dwarf – and that because such stars increase their luminosity over time, this main sequencer was about 5 billion years old, half-way through its 10-billion-year lifetime and among the most common planet-encircled star-type in the universe. From this they would surmise that the Earth is most probably only a bit younger than the Sun, the star-encircling planets of such systems having formed as a result of the merging-together "clean-up" of the gas and dust left over from star formation.

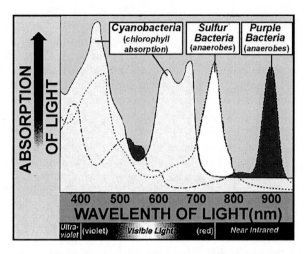

Fig. 9-6 Chlorophyll, the green pigment of cyanobacteria and higher plants, absorbs solar radiation in the visible portion of the Sun's spectrum as red and infrared wavelengths of the light pass though to power the photosynthesis of bacteria immediately below.

Earth-life's past history

Finally, from their collection of living specimens only – no fossils, no rocks to date, no knowledge Earth's geological history – what could the aliens deduce about Earth-life's past history and the sequential evolution of its various life-forms? Indeed, their planet-wide collection would lack any geologically based means to place whatever they interpret about the history of Earth-life into a reliable geochronological framework, the telling fossil evidence that led to Darwin's theory of biological evolution. Moreover, as they will have known from similar first-blush studies of other planets, even their educated guess about Earth's age (a bit younger than its companion 5 billion-year-old Main Sequence light-producer) would not be of much help, the age-related changes of the biotas of their previously sampled planets having varied widely – some changing rapidly, some more slowly – as a result of their varying planet-specific geologic histories.

Nevertheless, not all would be lost. Living organisms record the histories of their lineages in their morphologies and genetics. For example, bird wings and bat wings have a bone structure that evidences their common ancestry as the forelimbs of land vertebrates **(Fig. 1-7)**, and the bone anatomy of such vertebrates also shows their obvious evolutionary descent with modification from one major type to the next, a history recorded as well in their embryonic development **(Fig. 7-7)**. Yet the aliens would soon discover that the greatest treasure trove of evidence about the sequential development of Earth-life from its rudimentary beginnings resides in an organism's genes, a genetic underpinning that is most similar among closely related life forms but more disparate in those far separated by their histories, a lineage-relating comparison referred to as "phylogenetics."

As currently practiced by us Earthlings, phylogenetics has limitations. For example, given a sufficiently large set of specimens – such as the aliens' "Noah's Arc" – an amassed collection can reliably evidence the relative order of appearance and relative longevity of the various lineages. At the same time, however, this evidence is recorded in only the evolved and much curtailed set of genes that have survived to today. Thus, without the fossil record and the myriad extinct life-forms there preserved, the intermediate lineages which enormously outnumber the successful winners of the competition to survive, the aliens would be left to do little more than merely guess about precisely how one major group had given rise to the next.

Similarly, at least in our hands, phylogenetics reveals only the relative, not the geochronologic absolute ages of the various lineages. Because virtually all of the initial and intermediate members of extinct lineages have been lost from the living world, their no longer useful genes having been modified or deleted as well, phylogenetic "Trees of Life" almost always use the fossil record to place the beginnings of the lineages at times corresponding to their first known appearance in the geological record. Given the limited set of data thus available from studies of the genomics of extant life, Earthly phylogenetics has therefore not surprisingly devolved into a "numbers game," a simplified comparative statistical analysis of the more probable and less probable of the numerous phylogenic trees the data suggest may have actually occurred.

Of course, if you are creative you could easily imagine that the aliens have solved these problems and that they, unlike us, have devised a reliable "genetic clock," one that intuits the back-and-forth but now deleted underlying genetic changes from one lineage to the next and the time required for them to have occurred – the nitty-gritty of the evolutionary process. However, given what we currently understand about genetics – no doubt still rudimentary – and the well documented unpredictability of the timing and nature of evolutionary adaptations, such a solution seems unlikely.

Despite all this, the aliens could make broad phylogenetics-based close-to-the-mark deductions about **(7) how Earth's environment changed**

over time. Knowledge of changes in day-length would come from their understanding today's 24-hour circadian rhythm of Earth-life which would lead them to investigate its genetic underpinnings. Given that impetus, they presumably could then strip away the later time-adjusting changes the circadian module has undergone and from this reconstruct its original day-night cycle. Coupling this with their phylogenetic knowledge of the relative times of appearance of the various lineages, they would then discern that day-length has gradually increased over time. And from this and their knowledge of celestial mechanics and of the variation in a central planet's spin rate that results from the changing distance of an encircling satellite, they would know that the spatial relations of Earth and Moon have also changed over time.

Changes in Earth's atmospheric composition over the planet's past are also recorded in the genomics of Earth life – or, more accurately, in the metabolism- and biosynthesis-directing processes the genome generates. For Earthlings knowledgeable about Earth's geological record and the fossils there preserved, it is evident that the earliest atmosphere was essentially devoid of free oxygen, the O_2 in Earth's present-day atmosphere being a product of biological photosynthesis. Many of us also know that the first such oxygen-producers, about 3 billion years ago, were early-evolved microbial cyanobacteria and that for the following nearly one billion years the oxygen they produced was sponged up by its reaction with dissolved iron up-welled from oceanic basins **(Chapter 1)**. Over this enormously long episode, Earth's oceanic embayments and inland shallow seas "rusted," producing the ancient iron-rich ores on which the modern steel industry is based. As the deep-sea source of the iron ultimately became exhausted, the atmosphere became breathable by unicellular eukaryotes, precursors of life like us, a bit earlier than 2 billion years ago.

Because no such time-correlated evidence would be available to the aliens, they instead would turn to the genomics of their assemblage and the atmosphere-evidencing products they spawned. A survey of the life forms in their collection would show that aerobic respiration, the breathing process providing the cellular energy that sustains us humans and all other animals and plants, has three components. Glycolysis, the first part of the process, is anaerobic, meaning that it neither consumes nor produces oxygen. Overall, glycolysis is a straightforward ten-step series of reactions that by breaking down a molecule of the six-carbon sugar glucose into two molecules of the simple three-carbon biomolecule pyruvate produces a small amount of cellular energy. Like glycolysis, the second part of the process, the citric acid cycle, is also anaerobic and similarly produces only a small amount of energy. This cycle converts the glycolysis-generated pyruvate into forms useable by the next and final phase of the sequence, the electron transport system. It is only in this last phase of the process that oxygen plays a role, reacting with the products of the citric acid cycle to produce carbon dioxide, water and a large amount of cellular energy **(Chapter 8)**.

Thus, in all aerobic organisms the first two parts of the three-step process of breathing are anaerobic, in no way dependent on the presence of oxygen. From their studies of the bacteria in their collections, the aliens would then discover that both of the two non-oxygen using processes of the sequence play major roles in bacteria situated on the lower reaches of their phylogenetic tree. Glycolysis, the more primitive of the two, fuels the anaerobic bacterial heterotrophs near the base of the tree and the Citric Acid cycle (known also as the Krebs cycle) is an important component of anaerobic photosynthetic bacteria only a bit higher on the tree. Clearly, therefore, the oxygen-dependent breathing process of aerobes – the myriad occupants of the upper branches of their phylogenetic tree – must have developed in three distinct phases with the high energy generating use of oxygen being "added on" as a final step to two linked-together more primitive ways of living.

The aliens would also find that biosynthetic pathways, the way Earth-life produces particularly useful more complicated biomolecules from smaller simpler starting materials, tell the same story. Consider, for example, the synthesis of the carotenoid pigments that provide the bright red, yellow and orange colors of various vegetables and fruits, or the synthesis of the well-known mammalian sex hormones, testosterone and estrogen. In these pathways **(Fig. 8-6)** and a great many others, the use of oxygen has been added to a pathway that initially existed to produce other less complicated products.

By reasoning through their genome-based studies of the phylogenetics, metabolism and biosynthesis of their menagerie, the aliens would thus have telling insight into the history of the history of Earth's atmosphere. It would be apparent that in Earth's beginnings its atmosphere was virtually devoid of oxygen; that this gas, rather than issuing from the interior of the planet is a product of O_2-producing biological photosynthesis; and that the oxygen-dependence of the vast majority of life-forms in their collection evidences its late appearance but now long-term presence in the environment.

Knowing the history of atmospheric oxygen and that the Sun is a Main Sequence star, the aliens would also know of changes in Earth's ambient UV-flux, another aspect of the environment recorded in the genomes of the modern biota. Such stars have a predictable, easily discerned 10-billion-year life history from initially being less luminous, then slowly increasing to a stable brightness, and finally, as their interior nuclear reactions run out of fuel, expanding to become Red Giants and ultimately collapsing into White Dwarfs. In their initial one-to-two-billion-year-long less luminous phase of development, they emit large amounts of ultraviolet light that gradually decrease as the star's luminosity increases. Such UV-light is highly energetic and, thus, destructive to biomolecules, the stuff that makes up life. For early life, especially primitive photosynthesizers requiring sunlight to maintain their existence, this posed a huge problem, one they defeated by inventing UV-damage cellular repair mechanisms, many of which have been carried on to Earth's present-day life forms.

This history of the Earth's ambient UV-flux would be easy for the aliens to predict, but by investigating the UV-repair mechanisms encoded in the genomes of light-requiring and thus UV-exposed non-oxygen producing autotrophic microbes near the base of their phylogenetic tree – primitive anaerobic photosynthetic bacteria and their descendants, oxygen-producing cyanobacteria – they could confirm their supposition. As they might also do to unravel the history of circadian rhythms and the changes in Earth's day-length over the planet's history, they might strip away remnants of the intermediate changes in the UV-protective modules and reconstruct the original protective mechanisms. And, having their phylogenic data to order the sequence, they would then be able to compare the UV-sensitivity of the various mechanisms among the light-users lowest on their phylogenetic tree, anaerobic photosynthetic bacteria, with those of the most primitive oxygen-producing autotrophs, cyanobacteria, a bit higher on the tree, and among these to compare the lower- and higher-situated cyanobacterial groups. From such studies, their suppositions about the Sun, its history, and its influence on Earth life would be confirmed.

Having used phylogenetic genomic data to unravel the history of Earth's day-length, Earth-Moon relations, atmospheric composition, and ambient UV-flux, the aliens would then turn to the history of Earth's surface temperature. In the absence of the evidence recorded in the planet's rock record, this might not be quite as easy to figure out. But it already has been by us Earthlings – despite our still limited knowledge of the genetic make-up of the planet's biota – and if we could manage it, they could too, even faster and better.

Here's how we Earthlings solved the problem. Beginning some years ago I had been aware of oxygen isotope data from silica- (SiO_2-) rich minerals of ancient Earth rocks suggesting that the planet's photic zone temperature, the surface or near-surface where photosynthesis can occur, has decreased from about 167° F (75° C) to its present 59° F (15° F) over the past 3.5 billion years. Although this geochemistry-based finding was much disputed, the data being explained away as being due to heat- and pressure-related isotope changes, because I had provided the bulk of the samples used in the study (and knew from my studies of them that a great many had not been severely "pressure cooked") I imagined that the isotope-based interpretation might be correct. What was needed was an independent line of evidence.

Fortunately, I happened to know the world's expert in analyses of ancient temperature-indicating enzymes reconstructed from those of extant organisms (Akihiko Yamagishi of the Tokyo University of Pharmacy and Life Science). I therefore arranged for one of my graduate students to investigate the problem during two visits to his laboratory. These studies showed that the reconstructed ancient enzymes of modern photic-zone-inhabiting phototrophs (light-needing cyanobacteria, algae, and land plants) evidence the same marked global surface temperature decrease over geological time as that recorded in the rock record. Given this and the "Noah's Arc" of Earth-life amassed by the aliens, it seems certain that from such studies they, too, could uncover the marked changes in our planet's surface temperature that have occurred over its history.

The next on the aliens' now rapidly diminishing "to do list" would be to figure out **(8) how life began on Earth**. To us Earthlings, the sequence is simple – from simple to complex, first primitive anaerobic primordial soup-derived and soup-eating heterotrophs and only later sunlight-requiring autotrophs. To the aliens, this would also be easy to decipher.

Their earlier studies would have convinced them that the glycolysis-dependent microbes situated at the base of their phylogenic tree were among the first to appear. But what was the source of their glucose food? To the aliens the answer would be obvious, the non-biologic synthesis of simple organic materials which, as they would know well from their space-travels, is prevalent throughout the universe occurring in huge diffuse clouds of abiotic organics that span areas of interstellar space more extensive than our entire solar system.

The next major change in Earth-life, the appearance of anaerobic photoautotrophs, would also be easy to decipher. After all, such photosynthesizers make glucose rather than consume it from the environment. From this, the aliens would realize that the position of these first photosynthesizers on their "Tree of Life," a bit higher than the simpler glucose-consumers, meant that their development had freed Earth-life from its previous dependence on ready-made foodstuffs. Moreover, from their studies of the oxygen-producing photosynthesis and oxygen-consuming aerobic respiration of cyanobacteria even higher on their tree, the first on the tree to exhibit both of these traits, the aliens would also know that it was life-forms like these – in common parlance, "pond scum" – that set the stage for the thriving life of the present-day Earth.

By this stage in their quest, most of the pieces of the puzzle on their original "to-do list" would have fallen in place, ordered by the aliens' phylogenetic tree **(Fig. 8-7)**. Thus, they would be fully aware of the rise of complexity from heterotrophic anaerobes to non-oxygen-producing autotrophs to oxygen-producers and -users; from single-celled nucleated life to multicellular algae to marshland shrubs to upland forests **(Chapters 2, 3)**; from sponges to jellyfish to increasingly mobile diverse feeders **(Chapters 4, 5)**; and from fish to amphibians to reptiles to birds and mammals **(Chapter 6)**.

With this knowledge, an understanding of the aliens' final unsolved problem, **(9) how life changed over time**, would have come into focus and they would have finally managed to tackle all of the unknowns on their initial list of unanswered questions **(Fig. 9-7)**. Indeed, from comparison of the behavior of life forms in their menagerie, they might even be able to discern, or at least to make an educated guess about the history of intelligence, the ability of organisms to interact increasingly effectively with and manipulate their environment **(Fig. 8-12)**.

Yet from the aliens' study only of the living biota – no rocks or fossils, no clear-cut evidence of intermediate forms – they would remain stymied in their efforts to uncover exactly when and precisely how this sequential series of events actually occurred. Phased another way, the aliens would have good knowledge of the *products* of the evolutionary process, but no way to fully understand the timing and workings of the *process* itself. Interestingly, that is rather like the dilemma in which Charles Darwin found himself immersed upon return in 1836 from h
is round-the-world collecting trip on the *H.M.S. Beagle*, even though, thanks to the schooling by his geology mentor Charles Lyell he knew well the basics of Earth's most recent 500-million-year fossil record. From his collections during the sojourn, Darwin had myriad products, both living and fossil, of some wondrous process, but no clear knowledge of the underlying causative process itself, no way to firmly link products and process together.

ALIENS' FINAL LIST

CHARACTERISTICS OF EARTH-LIFE

Overall Composition

Two cell types	Initial biotic survey
Three Domains	Initial biotic survey

Overall Distribution

Latitudinal Variation	Initial biotic survey

Basic Structure

Hetero-Autotroph Dependence	Growth experiments

PHYSICAL PROPERTIES OF EARTH

Current Environment

Land-Sea Abundance	Collection map
Surface Temperature	Map plus growth experiments
Atmosphere Composition	Growth experiments
Atmosphere Pressure	Growth plus life maximum height

Global Properties

Earth Mass	Life maximum weight
Spheroid Shape	Autotroph distribution
Rotation Rate	Circadian clock
Earth Days/Months per Year	Growth layers

Astronomical Setting

Sun	Phototrophs
Moon	Monthly tides
Solar Flux	Chlorophyll absorbance
Age of Sun	Main Sequence star
Age of Earth	Typical solar system

EARTH-LIFE'S HISTORY

Changes in Earth's Environment

Day-length	Genetics of circadian clock
Earth-Moon Relations	Genetics of circadian clock
Atmosphere Composition	Metabolism and biosynthesis
Ultraviolet-Flux	Genetics of UV-protection
Surface Temperature	Genetics of autotrophs

Life's Beginnings

Simple to Complex	Metabolic pathways
Anaerobes First	Biosynthetic pathways

Life Over Time

Heterotrophs to Autotrophs	Prior findings plus genetics
Single- to Many-Celled	Prior findings plus genetics
Fish to Mammals	Prior findings plus anatomy

Fig. 9-7 The aliens' list of solved problems, based solely on their studies of extant Earth-life.

How did Darwin solve this dilemma? Spurred by the writings of T.R. Malthus **(Chapter 1)**, he made up an observation-driven thought experiment. He, like his contemporary, Alfred Russel Wallace **(Chapter 1)**, surmised, guessed, postulated that these products resulted from an innate urge for self-preservation, the need to compete, survive and succeed. From studies of the interactions among the diverse forms of life in their "Noah's Arc," the aliens might be able derive a similar process-based explanation of the products in their collection. But unlike the aliens, neither Darwin nor Wallace had access to such an extensive menagerie and neither had hardly any knowledge of the microbial world or even the slightest inkling of genes and phylogenetics.

What have we learned from this Thought Experiment?

This alien visitation thought experiment is of course a fanciful fable – it has not even a marginal real-world counterpart. As best we know, Earth has never been visited by aliens, benign and inquisitive or not. Nevertheless, it is good fun to imagine such stuff, to "think out of the box," and the basics of this tale well illustrate how science actually progresses. Darwin did such thought experiments, and Einstein and Oparin also. Now we have too.

As our thought experiment illustrates, quite often science sees the *products* of a process, but the underlying *process* that has given rise to the products is not readily apparent. Think about this in your own life as you trundle your way through the labyrinth of your own societal ecosystem, an ongoing competition to survive and thrive much like that in the natural ecosystem. You may sometimes feel stymied by your lack of progress and, since you don't understand the "why," the underlying process, imagine that your hoped-for success is plagued by some huge hoard of fellow competitors. In fact, however, their actual number pales into insignificance in comparison with the natural ecosystem, which is appreciably more enormous than you may have imagined. Indeed, according to the most recent, 2011, "best estimate," the total number of species on Earth numbers some 8.7 million of which 6.5 are on land, 2.2 million in the oceans. Note also that is only a rough guess, the same study calculating that the huge majority of living species, no less than 86% of all species on land and 91% of those in the seas, have yet to be discovered, described and catalogued. Moreover, each of us is a member of only one of the literally millions of species on Earth. Each of us is only a single tiny speck in the vast web of life.

Such caveats apply even more so to studies of the fossil record where they are magnified both by a lack of in-depth knowledge of the no-longer living products and the absence of direct evidence of the underlying product-producing process. This is easily understandable. Only members of a large successful group have a relatively "good" (but still only the tiniest, minuscule) chance of actually being fossilized in the rock record. Moreover, the fossil record, like the living world, provides direct evidence only of the *products* of the evolutionary process, not the *process* itself (the problem that plagued Darwin and Wallace). Nevertheless, and despite such difficulties, in the foregoing chapters of this book we have managed to pull together that limited fossil record to paint a convincing plant-first, animal-second, "eatees" followed by the "eaters" picture of how the Blooming of Life occurred.

How did we manage that? Spurred by understanding of the living plant and animal groups – spore plants and amphibian animals both requiring nearby water to reproduce, later-evolving seed plants and egg-laying reptiles freed from that requirement – we used the geological ages of the sequential events to place them in proper perspective, an understanding of life's past that well illustrates the contributions of fossil-finds spread across the globe. In particular, the colonization of land by primitive liverworts (close relatives of mosses) is first evidenced in the rock record 472 million years ago (Argentina), followed some 40 million years later, about 432 million years ago, by the earliest documented occurrence of *Cooksonia* (Czechoslovakia), the first land plant having a water-distributing central vascular core. The earliest known amphibian animal, *Elginerpeton* (Scotland), then followed this invasion of the marshlands about 368 million years ago. Soon thereafter, by about 360 million years ago, the plant lineage developed seeds that permitted their bearers to invade the highlands, the earliest known being the seed-fern *Elkinsia* (West Virginia, U.S.A.). The earliest known reptile, *Hylonomus* (New Brunswick, Canada), its egg-laying capability enabling it to follow suite, joined the upland invasion about 315 million years ago. Tiny shrew-like mammals (morganucodontids, Wales, 210 million years ago) and flowering plants (*Montsechia*, Spain, 130 million years ago) later first appeared, but by the time of their development, both plants and animals had mastered the problem of occupying the land surface. Note the numerous diverse localities of the tell-tale evidence – the unraveling of the history of life is the product of scientists worldwide!

The fossil-recorded products of this evolutionary sequence and their directly dated times of first appearance make good sense – land plants first, land animals later. Moreover, understanding of the self-preservation-spurred biological needs of plants and animals – most prominently, access to light and growing space for plants, and a reliable source food for animals – makes it easy to intuit the underlying the product-producing cause of these sequential developments with the nature of their habitat-adapting reproductive processes sealing the case.

This sequential co-evolution of plants and animals is thus a good example of use of the "use the *products* to infer the *process*" technique used by scientists to reveal reality. Nevertheless, because of the destruction of the aliens' relevant equipment, they would be "blind" to such geological age-data and thus hard-pressed to establish accurately the timing of this sequence of plant-animal interactions, whereas we can consult the fossil record and by using it as a guide have pieced the story together.

Like the thought experiments and logical strategies of Darwin, Einstein, and Oparin, our success in ferreting out this story of plant-animal sequential evolution well illustrates a basic underpinning of the scientific method. This aspect of how science works is perhaps even better illustrated by what I refer to as the "cascade of evidence approach," namely that if this is true, then this must be true, then this and this and this, each such finding supporting and reinforcing the ones before, step by step building confidence in the validity of the overall explanation **(Chapter 8)**. In essence, this technique is an example of what some teachers call "critical thinking" and philosophers refer to as the Socratic Method, a matter of interrogating, checking and confirming the observable results in an effort to uncover their underlying cause.

Scientists carry out this exercise day-in, day-out, testing the data known to them and uniting them, via a cascade of evidence, into an all-encompassing paradigm – an underlying fundamental aspect of the broad-brush workings of science masterfully elucidated by philosopher and historian of science Thomas S. Kuhn **(Chapter 8)**. It's fair to ask why scientists go through all this fuss. Why don't they simply state their conclusion, leave it at that, and move on? The answer is simple. They know that there is only one reality, only one real world, and their task is to get the story straight. A "guess-or-by-golly," good enough is "good enough" description of that world would not do the trick. Indeed, their overriding mantra, codified by Nobel Laureate physicist Richard P. Feynman (1918-1988) in his 1974 Cal Tech commencement address, is that *"You must not fool yourself – and you are the easiest person for you to fool!"*

As a final exemplar of this approach, let's recall what the aliens learned about Earth and its life and how they figured it out. They began their quest by examining the basic characteristics of Earth-life, the composition, distribution and structure of the planet's biota. From this cursory beginning, they realized that Earth-life would have to fit the characteristics of the planet itself so they then investigated Earth's physical properties, its environment, global properties and astronomical setting. With these in hand they could then finally ask about the history of the Earth, how its environment changed over time, how primordial life arose from its simple nonbiologic chemical precursors, and how life then evolved from such humble beginnings into the myriad life-forms that now populate the globe. The step-by-step workings of the aliens' Socratic cascade of evidence are summarized in **Fig. 9-8**. From their experience and knowledge – perhaps by applying their own evolutionary past to analyzing the Earth – the aliens amassed the products, surveyed their varied traits, analyzed the life forms to reveal their environment, used these to place Earth in its celestial setting, and then put the package together to understand the history of Earth, its environment and its biota. Quite an accomplishment!

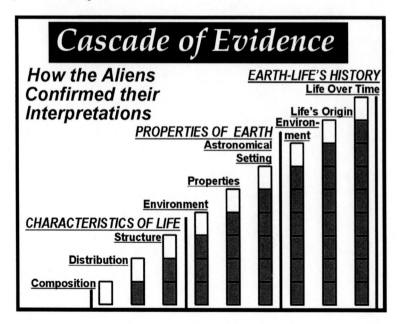

Fig. 9-8 The cascade of evidence amassed by the aliens in their first-blush analysis of Earth-life.

Ever since Darwin, we Earthlings have performed this task as well. Hark back to what the American-Russian geneticist Theodosius Dobzhansky (1900-1975) taught us in 1973 **(Preface and Chapter 1)**: *"Nothing in biology makes sense except in the light of evolution."* Darwin, Dobzhansky, and the myriad other scholars of biological evolution are correct – evolution *is* the key to unraveling the interrelated history of life and its changing environment. This great truth provides the underpinning of the prime take-home-lesson from the *Blooming of Life,* namely, as we now know and the visiting aliens would have learned as well, Earth-life and its environment are intimately intermeshed – from bacteria to bats, plants to people, microbes to man, ***all life is linked!***

GLOSSARY OF TECHNICAL TERMS

Acoelomate .... metazoan lacking a coelom
Acritarch............. fossil eukaryotic phytoplankton, single-celled microalgae
Actiniaria.. coral-like sea anemones
Actinopterygian .... ray-finned fish
Adenosine triphosphate (ATP)........... a biochemical containing high-energy phosphate bonds
Aerobic respiration........ oxidation ("burning") of foodstuffs to yield energy
Aerobic .... pertaining to the use or presence of molecular oxygen
Age of Dinosaurs ....informal name for the Mesozoic Era
Age of Discovery ....the 15th to 18th centuries of European history
Age of Flowering Plants....................... informal name for the Cenozoic Era
Age of Invertebrates and Fish ....informal name for the Paleozoic Era
Age of Mammals.................................. informal name for the Cenozoic Era
Age of Naked Seed Plants..................... informal name for the Mesozoic Era
Age of Spore Plants ....informal name for the Paleozoic Era
Agnatha .... primitive jawless fishlike vertebrate animals
Alpha Centauri .... closest star system to Earth's Solar System
Ammonia... hydrogenated nitrogen, NH3
Amniote egg.................. hard-shelled egg of reptiles, birds and monotremes
Amniote................vertebrates having an embryo- or fetus-enclosing amnion
Amphibia .... vertebrates having aquatic larvae and air-breathing adults
Anaerobic .... pertaining to the absence of molecular oxygen
Analogous traits .... similar traits of differing evolutionary derivation
Annelida .... taxonomic family of segmented worms
Archaea .... archaeal microbes, members of 1 of 3 Domains of Life
Arthropoda ....invertebrate animals having jointed limbs
Asexual reproduction .... binary cell division, cloning of the parent cell
Asteroidea................................class of echinoderms that includes sea stars
Autotrophy............"self-feeding" lifestyle of plants and plant-like microbes
Aves ....class of vertebrates composed of birds

Bacteria .... bacterial microbes, members of 1 of 3 Domains of Life
Banded iron formation ...."BIF," a type of sedimentary iron-rich rock
BIF...................banded iron formation, a type of sedimentary iron-rich rock
Bilateria................. animals having bilateral symmetry at some stage of life

Biogenic carbon .... carbon of biologic origin
Bitter Springs Formation .... Precambrian 850 Ma fossiliferous unit
of central Australia
Blastopore .... hole in blastula that becomes either the anus or the mouth
Blastula .... ball of cells formed early in animal embryos
Brachiopod .... lophophorate shelled invertebrate animals
Bryozoan .... lophophorate colonial invertebrate animals
Burgess Shale .... fossiliferous 510 Ma geological unit of Alberta Canada

Calcichordate .... Paleozoic fossils thought to be related to extinct
echinoderms
Cambrian .... Paleozoic Geological Period 541-485 Ma
Cambrian Explosion of Life .... marked increase in types of animals
during the Cambrian
Carbonate .... CO3-minerals or the rocks (e.g., limestones) they compose
Carboniferous .... Paleozoic Geological Period 360-300 Ma
Cascade of Evidence Test .... independent lines of hypothesis-confirming
evidence
Cenozoic Era .... 65 Ma ago to present
Cephalochordata .... chordate animals commonly called amphioxus
or lancelets
CH2O .... chemical formula for glucose sugar, C6H12O6
Changzhougou Formation .... fossiliferous 1,800 Ma geological unit of
North China
Chicxulub Impact Event .... meteor impact in causing the K-T extinction
Chloroplast .... organelle for oxygenic photosynthesis in eukaryotic cells
CHON(SP) .... carbon, hydrogen, oxygen, nitrogen (sulfur, phosphorus)
Chordata .... sea squirts, lancelets and vertebrate chordate animals
Chromosome .... gene- (DNA-) containing elongate bodies in cell nuclei
Chuar Group .... 800 to 742 Ma fossiliferous geological unit of Arizona
Citric acid cycle .... the cyclic part of the aerobic respiration pathway
Coelacanth .... a fish having a three-lobed tail fin and fleshy pectoral fins
Coelenterate .... invertebrate animal such as jellyfish and corals
Coelom .... body cavity between the body wall and the digestive tract
Coelomate .... metazoan having a coelom
Concentricycloidea .... sea daisies, a type of echinoderm
Condrichthyes .... cartilaginous fish
Cretaceous .... Mesozoic Geological Period 150-65 Ma
Crinoidea .... class of echinoderms that includes crinoids
Crossopterygii .... lobe-finned fish regarded as ancestral to amphibians
Cutin .... waxy coating on stems and spores that inhibits desiccation

Cyanobacteria .... O2-producing and consuming bacterial prokaryotes
Cycadales .... gymnosperms having a short trunk and pinnate leaves
Cyclostomata .. agnathan chordates, the living jawless lampreys and hagfish

Deoxyribonucleic acid (DNA) .... gene-encoding biochemical of chromosomes
Deuterostome .... animals in which the blastopore becomes the adult anus
Devonian .... Paleozoic Geological Period 420-360 Ma
Dicot .... type of flowering plant having net-veined leaves
Dicotyledonous angiosperm .... flowering plant having two seed leaves
Dinosauria .... extinct reptiles widely distributed during Mesozoic
Diploid .... "2N," twice the number of chromosomes of haploid gametes
DNA .... deoxyribonucleic acid, the genetic material of chromosomes
Domains of Life .... the Super-kingdom-like groups Archaea, Bacteria
and Eucarya
Doushantuo Formation .... Ediacaran 600 Ma fossiliferous unit
of southwestern China

Ecdysozoa .... rRNA-based subdivision of protostome animals
Echinoderm ...... invertebrate such as a sea star, sea urchin, or sea cucumber
Echinoidea .... echinoderm class that includes sea urchins and sand dollars
Ectoderm .... outermost tissue layer
Ectothermy .... body heat absorbed from the environment
Ediacaran .... Precambrian Geological Period 600-541 Ma
Endoderm .... innermost tissue layer
Endosymbiosis .... the living of one kind of organism within the cells
of another
Endothermy .... body heat generated by internal metabolism
Enteropneusta .... hemichordate worms such as *Balanoglossus*
and related genera
Enzyme ..... biochemical that governs and speeds chemical reactions in cells
Epidermis .... outermost cell layer, "skin"
Eucarya .... eukaryotes, members of 1 of 3 rRNA Domains of Life
Eukaryote .. organism (protist, plant, fungus, animal) having nucleated cells
Eutherian .... placental mammals of the major chordate group Eutheria

First Law of Biology .... to stay alive
Formicidae .... family of insects that includes all ants

Ga .... Giga Anna, one billion years (1 x 10^9 yr)
Gamete .... haploid egg or sperm
Geological Period .... formal division of geological time

Ginkgoales........Mesozoic gymnosperms having only one surviving species
Global Warming............temperature rise attributed to the greenhouse effect of CO_2
Glucose synthesis ....metabolic production of glucose sugar
Glucose.. the six-carbon sugar $C_6H_{12}O_6$
Glycolysis .... anaerobic breakdown of glucose sugar to pyruvate
Gramineae.................................. the taxonomic family of monocot grasses
Green River Formation .... Paleogene (Eocene, 15 Ma) unit of southwestern Wyoming
Gunflint Formation.......... fossiliferous 1,900 Ma geological unit of Ontario Canada

Haploid................... "1N," half the number of chromosomes in diploid cells
Hemichordate Hypothesis .... accepted theory linking echinoderms and chordates
Hemichordate.......... wormlike chordates having gill slits and a nerve chord
Heterotrophy .... "eating others" lifestyle of animals and many microbes
Holoblastic ....spiral cleavage pattern of protostome blastulas
Holothuroidea ..... a class of echinoderms that comprises the sea cucumbers
Homologous traits .... traits having a shared evolutionary derivation
Hox genes ....DNA-genes specifying head-to-tail animal body plans

Insecta ....small arthropods having six legs and pairs of wings
Interbreed.................................... able to mate and produce viable offspring

Jurassic...Mesozoic Geological Period 205-150 Ma

K-T extinction.. major extinction event at the Cretaceous-Tertiary boundary
Kuhn Cycle .... an all-encompassing synthesis of how science advances

Lakhanda Formation........ 1,020 Ma geological unit of southeastern Siberia
Law of Unintended Consequences......... undesired outcome of a well-meant action
Lobe-finned fish........... fish having muscular fins, precursors of amphibians
Lophophore .... arm-like feeding structure of bryozoans and brachiopods
Lycophyta .... plants having spore cones and helically arranged leaves

Ma .... Mega Anna, one million years (1×10^6 yr)
Magnoliales....................................early-evolving order of flowering plants
Mammalia........class of warm-blooded vertebrates having mammary glands
Marsupial .... mammal having offspring suckled in a belly-pouch

Meiosis .... cell division that produces plant spores or animal gametes
Mesoderm .... middle tissue layer
Mesozoic Era .... 250 to 65 Ma ago
Messenger ribonucleic acid (mRNA) ..... RNA carrying the genetic message
from DNA to ribosomes
Metabolism .... the buildup or breakdown of biochemicals in cells
Microbe .... prokaryotic microorganism, either bacterial or archaeal
Microfossil .... minute fossil, typically 100 μm or smaller
Mitochondrion .... organelle of aerobic respiration in eukaryotic cells
Mitosis .... eukaryotic body cell division that makes exact copies, "clones"
Mollusca .... invertebrates having a soft body protected by a calcareous shell
Monocot .... type of flowering plant having parallel-veined leaves
Monocotyledonous angiosperm .... flowering plant having one seed leaf
Monomers .... small "building block" compounds that link to compose
polymers
Monotreme egg-laying mammal having one opening for excretion and birth
mRNA .... messenger ribonucleic acid
Mutation .... change in the genetic material (DNA) of cells

N₂-fixation ....metabolic conversion of atmospheric nitrogen to ammonia
Natural Selection ...... organisms better adapted to their environment survive
Neogene .... Cenozoic Geological Period 25 Ma-present
Notochord .... lengthwise flexible rod in vertebrate embryos

Ophiuroidea .... class of echinoderms that includes brittle stars
Ordovician .... Paleozoic Geological Period 485-445 Ma
Ornithischia .... herbivorous dinosaurs having a bird-like pelvic structure
Osteichthyes .... bony fish
Oxidized iron .... iron $^{+}3$, as in the iron oxide mineral hematite, $Fe2^{+3}O3^{-2}$
Oxygen sink ...... substance (volcanic gases, reduced Fe) that sponges up O2
Oxygen-producing photosynthesis .... O2- and sugar-producing metabolic
process

Paleogene .... Cenozoic Geological Period 65-25 Ma
Paleozoic Era .... 550 to 250 Ma ago
Pelycosaur .... primitive mammal-like reptiles
Pentameral .... five-fold symmetry
Permian .... Paleozoic Geological Period 300-250 Ma
Phanerozoic Eon .... 541 Ma ago to present
Phloem .... thin-walled food conducting tubes in plants
Photoautotrophy .... autotrophy powered by light energy

Photosynthesis... photoautotrophy
Placental........................... mammals having a placenta, eutherian mammals
Placodermi..... class of extinct armored fish having primitive jaw structures
Platyhelminthes ....phylum of bilaterian flatworms having a soft body
Plesiosaur........... large marine Mesozoic reptiles having paddle-like limbs
Polymer .... long chemical compounds composed of monomeric subunits
Polypodiophyta... fern and fern allies spore plants
Population .... organisms of the same type in a particular region
Posidonia Shale........ Jurassic 150 Ma geological unit of southern Germany
Pound Quartzite ..... fossiliferous Ediacaran 600 Ma unit of South Australia
Precambrian Eon .... formation of Earth ~4,500 Ma ago to 541 Ma ago
Principle of Faunal Succession .... fossil assemblages systematically vary
                                                                                              older to younger
Prokaryote............... microbes (Bacterium, Archaean) having cells that lack
                                                                                                        a nucleus
Protein.. polymer composed of amino acids
Protostome .... animals in which the blastopore becomes the adult mouth
Proxima Centauri............ nearest star to the sun, part of a triple-star system
Pterobranchia... small worm-shaped deep-sea tube-dwelling hemichordates
Pterosaur........................... extinct flying reptiles having featherless wings
Pyruvate .... salt of the simple 3-carbon compound pyruvic acid, C3H4O3

Reduced iron............... iron $^{+2}$, as in the iron sulfide mineral pyrite, $Fe^{+2}S2^{-1}$
Reproductive isolation............ geographic or other isolation mechanism that
                                                                                                prohibits mating
Reptilia .... class of cold-blooded air-breathing bony vertebrates
Rhizome .... underground stem
Rhynie Chert............ fossiliferous 410 Ma geological unit of Aberdeenshire
                                                                                                        Scotland
Rhyniophytina............ plants having naked dichotomizing axes and terminal
                                                                                                        spore sacs
Ribosome .... minute intracellular structures that produce proteins
Root .... plant anchoring organ
rRNA.. RNA of protein-making ribosomes

Saurischia.. dinosaurs having a lizard-like pelvis
Sauropod .... large extinct herbivorous tetrapod saurischian dinosaur
Second Law of Biology... to mate and reproduce
Sequential co-evolution .... interrelated sequential evolution of two types
                                                                                                        of organisms

Sexual reproduction .... combination of genes from genetically differing parents
Silurian ....Paleozoic Geological Period 445-420 Ma
Solnhofen Limestone.......... fossiliferous 155 Ma geological unit of southern Germany
Spearfish Formation.... Triassic 250-205 Ma geological unit of northeastern Wyoming
Sphenophytina .... plants having terminal spore cones and whorls of microphylls
Spiralia ....rRNA-based subdivision of protostome animals
Sporangium .... sac in spore plants where spores are produced by meiosis
Spore.. haploid ("1N") products of meiosis in plants
Stomate .... pores in plant epidermis for absorption of carbon dioxide
Stromatolite ..... layered mound-like masses built by microbial communities
Superposition...........................overlying in a sequence of sedimentary rocks
Synapsida ....proto-mammals superficially similar to reptiles
Syngamy .... fusion of gametes during sexual reproduction

Telome Theory............. explanation for the development of fern megaphylls
Tertiary...........alternative name for the Cenozoic (Paleogene and Neogene)
Tetrapod .... four-legged vertebrate
Theory of Special Relativity............... energy and mass are interchangeable depending on velocity
Triassic ....Mesozoic Geological Period 250-205 Ma
Tube feet .... water-filled tubular suction cup-like "feet" of echinoderms
Tunicata...............................chordate animals such as sea squirts and salps

Valve.................... either of the paired shells of brachiopods and pelecypods
Vertebrata.......... chordates having a segmented usually bony spinal column

Xylem ....thick-walled water conducting tubes of vascular plants

Yixian Formation... Cretaceous 125 Ma geologic unit of northeastern China

Zosterophyllophytina................. extinct naked or spiny plants having lateral sporangia

INDEX

A History of Earth's Biota: The Blooming of Life   259

Curie, Pierre, 10
Cuvier, Baron Georges, 18, 19,
    193
Cyanobacteria, 3-5, 32, 42, 89,
    105, 233, 234, 238,
    239; cellular
    components of, 5;
    global success of, 4;
    mat-communities of,
    233; photosynthesis
    of, 4, 238; non-sexual
    reproduction of, 5
Cycadales, 76
Cycas revoluta, 76. 77
Cyclostomata, 160
Czar Nicholas II, 38

Daeschler, Edward, B. ("Ted"),
    165
Dalí, Salvador, 218
Darroch, S. A. F., 138
Darwin, Charles Robert, x, xiv,
    11, 12 18, 20, 36, 80,
    141, 144, 192, 193,
    240. 242-244, 246;
    beetle collections of,
    144, 145; and the
    missing pre-Cambrian
    fossil record, x, 196;
    and *On the Origin of
    Species*, x, 18, 20, 93,
    195, 196; and origin
    of life, 36; and Theory
    of Evolution,11, 36
Darwin, Erasmus, 18, 193
Darwin's Beetle, 145
Darwin's Finches, 193
Darwinian Evolution, xiv, 16,
    18, 21, 24, 36, 37,
    192, 195, 201, 220,
    234; compared to
    American history, 23;
    and Natural Selection,
    193, 201, 203;
    precursors of, 18;
    rules of, 21, 23;

simple definition of,
    21, 22; solution to
    anomaly of, 200; as
    test case of Kuhn
    Cycle, 192-205
de Vries, Hugo, 196
Degeneria, 79
DeMott, Larry, x; and the
    greatest unsolved
    problem in natural
    science, x
Dendrostomum, 132
Deuterostome, 17, 104, 121,
    146-149, 154, 157,
    158, 164, 185;
    linguistic derivation
    of name, 147;
    meaning of name, 147
Devonian Period, 12, 13, 42,85,
    110, 128, 146, 161-
    164, 167, 209, 231;
    duration of, 13;
    geographic basis of
    name, 13
Dickinsonia, 98
Digitalis purpurea, 35;
    treatment for heart
    disease,35
Dilcher, David, 80
Dimetrodon, 177
Dinichthys, 161
Dinopedia, 137
Dinosauria, 1, 2, 171, 172, 177-
    180, 185;
    characteristics of,
    171; linguistic
    derivation of name,
    171
Djarthia, 182
Dobzhansky, Theodosius
    Grygorovych, xiv, 15, 16,
    246; quotation of, xiv, 15,
    246
Domains of Life, 32, 204, 226,
    227
Doushantuo Formation, 102